Piston Engine-Based Power Plants

The Power Generation Series

Paul Breeze—Coal-Fired Generation, ISBN 13: 9780128040065
Paul Breeze—Gas-Turbine Fired Generation, ISBN 13: 9780128040058
Paul Breeze—Solar Power Generation, ISBN 13: 9780128040041
Paul Breeze—Wind Power Generation, ISBN 13: 9780128040386
Paul Breeze—Fuel Cells, ISBN 13: 9780081010396
Paul Breeze—Energy from Waste, ISBN 13: 9780081010426
Paul Breeze—Nuclear Power, ISBN 13: 9780081010433
Paul Breeze—Electricity Generation and the Environment, ISBN 13: 9780081010440

Piston Engine-Based Power Plants

Paul Breeze

ACADEMIC PRESS

An imprint of Elsevier

Academic Press is an imprint of Elsevier
125 London Wall, London EC2Y 5AS, United Kingdom
525 B Street, Suite 1800, San Diego, CA 92101-4495, United States
50 Hampshire Street, 5th Floor, Cambridge, MA 02139, United States
The Boulevard, Langford Lane, Kidlington, Oxford OX5 1GB, United Kingdom

Notices
Knowledge and best practice in this field are constantly changing. As new research and experience broaden our
understanding, changes in research methods, professional practices, or medical treatment may become
necessary.

Practitioners and researchers must always rely on their own experience and knowledge in evaluating and using
any information, methods, compounds, or experiments described herein. In using such information or methods
they should be mindful of their own safety and the safety of others, including parties for whom they have a
professional responsibility.

To the fullest extent of the law, neither the Publisher nor the authors, contributors, or editors, assume any
liability for any injury and/or damage to persons or property as a matter of products liability, negligence or
otherwise, or from any use or operation of any methods, products, instructions, or ideas contained in the
material herein.

British Library Cataloguing-in-Publication Data
A catalogue record for this book is available from the British Library

Library of Congress Cataloging-in-Publication Data
A catalog record for this book is available from the Library of Congress

ISBN: 978-0-12-812904-3

For Information on all Academic Press publications
visit our website at https://www.elsevier.com/books-and-journals

  Working together
to grow libraries in
developing countries

www.elsevier.com • www.bookaid.org

Publisher: Joe Hayton
Acquisition Editor: Maria Convey
Editorial Project Manager: Mariana Kuhl
Production Project Manager: Vijayaraj Purushothaman
Cover Designer: MPS

Typeset by MPS Limited, Chennai, India

CONTENTS

CHAPTER 1

An Introduction to Piston Engine Power Plants

Piston engines or reciprocating engines (the two terms are often used interchangeably to describe these engines) are by a wide margin the largest group of thermodynamic heat engines in use around the world. Their applications range from model aeroplanes to lawn mowers: they include all the automotive power plants found in motor cycles, cars, trucks and many other sorts of heavy machinery; they power locomotives, ships and many small aircraft and they provide stationary electrical power and combined heat and power to numerous sites across the globe.

The number in use is enormous; the United States alone produces more than 35 million each year. Engines vary in size from less than 1 kW (model engines can be a few watts) to 80,000 kW. They can be driven using a wide range of fuels including natural gas, biogas, liquefied petroleum gas, gasoline, diesel, bio-diesel, heavy fuel oil and even coal. They are manufactured all over the globe and there is a large global base of expertise in their maintenance and repair. While modern engines are often extremely advanced and digitally controlled, older engines can often be kept in service by small, local workshops.

In line with the wide range of engines available, the power generation applications of piston engines are enormously varied. Small units can be used for standby power or for combined heat and power in homes and offices. Larger standby units are often used in situations where a continuous supply of power is critical such as in hospitals or to support highly sensitive computer installations like an air traffic control system or one of the many computer server farms around the world. Commercial and industrial facilities use medium-sized piston engine-based combined heat and power units for base-load, distributed power generation. Large engines, meanwhile, can be used for base-load, grid-connected power generation while smaller units form one of the main sources of base-load power to isolated communities with no access to an electricity grid.

Piston Engine-Based Power Plants. DOI: https://doi.org/10.1016/B978-0-12-812904-3.00001-X

The piston engines used for power generation are almost exclusively derived from similar engines designed for motive applications. Smaller units are normally based on car or truck engines while the larger engines are based on locomotive or marine engines. Performance of these engines vary. The small engines are usually cheap because they are mass produced but they have relatively low efficiencies and short lives. Larger engines tend to be more expensive but they will operate for much longer. Large, megawatt-scale engines are among the most efficient prime movers available,[1] with simple cycle efficiencies approaching 50%.

The piston engine takes its name for the characteristic feature of the engine design, a piston. This piston moves backwards and forwards (or up and down) within a cylinder, that is sealed at one end, in response to the expansion and contraction of a gas within the sealed chamber as the gas is heated and cooled. The heating and cooling of the gas sealed in the piston cylinder can be carried out by applying alternate heating and cooling externally, in which case the engine is called an external combustion engine. However in most engines of this type the heating takes place via the combustion of a fuel in air inside the cylinder itself. This type of engine is called an internal combustion engine.

There are two principle types of internal combustion reciprocating engines, the spark ignition engine and the compression or diesel engine. The latter was traditionally the most popular for power generation applications because of its higher efficiency. However it also produces high levels of atmospheric pollution, particularly nitrogen oxides. As a consequence spark ignition engines burning natural gas have become increasingly popular units for power generation, at least within industrialised nations. A third type of piston engine, called the Stirling engine, is also being developed for some specialised power generation applications. This engine is novel because it is an external combustion engine.

The enduring popularity of the piston engine derives from it portability and flexibility. With a small reservoir of fuel the engine is self-contained and can produce energy or work for extended periods. This

---

[1]Slow-speed engines are the most efficient engines for converting fuel energy via heat into rotary motion to generate electricity. Fuel cells, which turn chemical energy directly into electrical energy, can be more efficient.

is particularly important for transportation applications such as road vehicles but also makes them useful for the wide range of power generation applications noted above.

The main drawback of these engines is that they generally require the combustion of a fossil fuel to provide the energy needed to drive them. In consequence they represent a major source of atmospheric carbon dioxide emissions across the globe and contribute significantly to the quantity of this gas that is released into the atmosphere each year. In addition both types of internal combustion engine, but particularly diesel engines, are sources of a range of other pollutants. Where these engines are used for stationary applications such as power generation it is feasible to apply advanced techniques to clean the exhaust gases and reduce their atmospheric emissions of carbon dioxide. However this is not cost effective for smaller, mobile-scale applications such as for cars. In consequence there is a major industrial effort taking place to find a cleaner replacement for transportation applications. Candidates include fuel cell-powered vehicles and battery-powered vehicles.

THE HISTORY OF THE PISTON ENGINE

The earliest references to the concept of a piston engine can be found in the 17th century when the French inventor Jean de Hautefeuille proposed a device that would use gunpowder as the fuel to drive a piston in a cylinder as a means of generating mechanical energy. This device operated using single charges of gunpowder and would have had to be recharged before each cycle so would have been of limited use as an engine for providing useful work. There is no evidence that Hautefeuille actually built his device but the Netherlands' scientist Christiaan Huygens may have attempted to do so. Material limitations would, anyway, have made it difficult to develop an engine using this principle at that time. Nevertheless, Hautefeuille's proposal appears to have been the first mention of both a piston engine and the idea of an internal combustion engine.

While the materials were not available to build successful engines of this type, another proposal from the 17th century did gain traction, an external combustion engine using steam as the working fluid. One of the earliest contributors to this line of development was another French scientist, Denis Papin. He put forward the idea of using steam as a working fluid to generate a pressure or force to drive a piston through one stroke of a cycle, with cooling water used to condense the steam again during the second stroke of the cycle when atmospheric pressure would return the piston to its starting point. Again, Papin does not seem to have developed a practical engine based on this design but another inventor did, Englishman Thomas Newcomen, who in 1712 published his design for an atmospheric engine, so called because one side of the cylinder is open to the atmosphere, for use as a pump. This device is now commonly called the Newcomen engine (Fig. 1.1).

The Newcomen engine was designed as a pump to remove water from underground mine workings. The device consisted of a large cylinder into which a piston was inserted from above, sealing the cylinder from the top but with the top of the piston open to the atmosphere. The piston was connected via a rod to a beam that operated through a

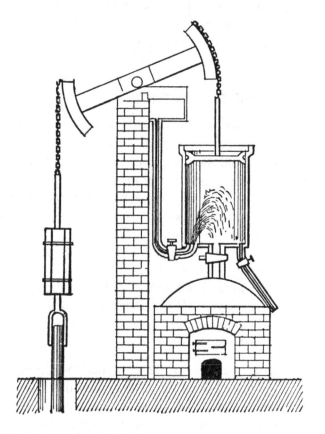

Figure 1.1 The Newcomen engine. Source: Wikipedia.

pivot to raise and lower a second rod which drove a pump in the mine, below. The bottom of the cylinder was closed but was connected through a valve to a boiler that produced very low-pressure steam. The boiler for the engine was typically a 'haystack' boiler, so called because of its shape, which could produce steam to a maximum of 0.2–0.3 bar above atmospheric pressure. During the first part of the engine cycle this steam was admitted into the cylinder, forcing the piston to rise against the pressure of the atmosphere on the top of the piston. The valve was then closed and a second valve opened briefly, allowing cold water to spray into the cylinder and cool this steam which then condensed, allowing atmospheric pressure to force the piston back down. This cycle was repeated around 12 times each minute.

Operation of the engine required the two valves to be alternately opened and closed, with this cycle of opening and closing repeated at regular intervals. The valves were originally operated manually but a mechanical means of carrying this out was soon devised, an early forerunner of the valve mechanisms in modern engines.

The development of steam piston engines continued through the 18th century and into the 19th century with designers such as James Watt, who developed a more efficient version of the Newcomen engine. All the early steam engines used low-pressure steam but gradually, as material technology improved, higher pressure was introduced.

The use of high-pressure steam allowed smaller engines to be built. Piston steam engine development continued until the late 19th century when the steam turbine first appeared. This was a much more advanced, high-speed engine and it quickly superseded steam reciprocating engines for most applications. However the use of reciprocating engines for steam locomotives on railway systems continued, in some regions, until the end of the 20th century.

While steam, external combustion, reciprocating engines played an important role in the development of piston engines it is the internal combustion engine that has proved to be the most enduring engine of this type. Although the concept was proposed in the 17th century, internal combustion was virtually abandoned when steam engine development began and it was not until the 19th century that practical engines of this type began to reappear.

One of the first was designed by Belgian engineer Jean Joseph Etienne Lenoir. His single cylinder engine was an adapted steam reciprocating engine that was fired with fuel gas that was ignited inside the cylinder using 'jumping sparks', an early electrical ignition system. Lenoir's engine was derived from a double-acting steam engine in which steam is used to drive the piston in both directions. (In the Newcomen engine steam drives it in one direction and atmospheric pressure in the other.) Lenoir engines were primarily used for stationary applications such as pumping but Lenoir did also develop carriages driven by his engine, some using liquid hydrocarbon fuel.

The Lenoir engine was primitive and noisy but it did work and it inspired the German engineer Nikolaus August Otto, who together with his brother built a copy of the Lenoir engine in 1861. The brothers tried to obtain a patent for their engine in Prussia, but failed. The same year they experimented with a charge of fuel and air that was compressed instead of at atmospheric pressure. Although this compressed charge engine only ran for a short time before failing, this idea was to prove an essential component of the internal combustion engine.

By 1864 Otto had set up a partnership with Eugen Langen to develop engines. The first Langen Otto engine was an atmospheric engine that used internal combustion to drive the piston through the first part of the cycle while the force of atmospheric pressure returned the piston as in the Newcomen engine. This proved a successful internal combustion engine but it developed only low power and required significant head room to operate.

Otto continued to work on improvements and in 1876 the company, now called Deutz AG Gasmotorenfabrik, developed the four-stroke, compressed charge engine based on an engine cycle that is now known as the Otto cycle. This forms the basis for all the spark ignition engines that are in use today. One of the main differences between this and earlier engines was that it did not rely on atmospheric pressure to return the cylinder after the power stroke. Instead a large flywheel on the power shaft was used to provide sufficient angular momentum to complete the cycle. The engine was fuelled with coal gas and used a live flame to ignite the fuel–air mixture at the top of the compression stroke of the engine.

Other internal combustion engine cycles were also being pursued. Two men, German engineer, Karl Benz and an Englishman, Dugald Clark, independently developed two stroke engines in 1878. Benz received a patent for his design in Germany in 1879 while Clark obtained a patent in England in 1881. Benz went on to develop and patent spark plugs, the carburettor and the clutch and gear shift.

The other main branch of internal combustion reciprocating engines was also launched by a German engineer, Rudolf Diesel. He had studied thermodynamics closely and was seeking a more efficient engine. This led him to the design of the compression ignition engine in 1892 and a patent for the design in 1893. The principle difference between this and Otto's engine was that instead of an air–fuel mixture being admitted into the cylinder before compression, only air was admitted and it was compressed more highly than in the Otto engine. This elevated the temperature of the air to a point at which fuel, when admitted, ignited spontaneously. The diesel engine has become widely used in many applications because of its greater fuel efficiency. However the engine produces more polluting emissions than the Otto cycle engine and widespread use is being questioned today, particularly for road vehicles.

Since the pioneering work of the late 19th century there have been major refinements to both Otto cycle and Diesel engine design. These include turbo charging or supercharging, the introduction of advanced electronic ignition systems and a range of different engine configurations. There have also been attempts at different designs. During the early years of the 20th a rotary engine, in which a circular arrangement of cylinders rotated around a stationary crankshaft became popular, particularly for aircraft. However inherent limitations led designers back to the traditional reciprocating design as this advanced. A completely different rotary engine, the Wankel engine, was developed later in the century but has also found only limited application.

GLOBAL RECIPROCATING ENGINE, POWER GENERATION INSTALLATIONS

The application of reciprocating engines for power generation is so large and diverse that it is impossible to gauge accurately the total installed capacity of this type of generator, either by country or globally. The range of stationary applications of this type includes base-load power plants based on large, low-speed diesel engines and upon medium- and low-speed gas engines. A large number of both diesel and gas engines are used for distributed generation where they supply base load and peak power directly to consumers. Similar systems are frequently used to supply power to consumers that are not connected to the grid. In recent years a range of hybrid systems that employ a mix of renewable energy, usually from wind or solar generators, energy storage and reciprocating engines have become popular in both distributed generation and remote power applications.

Another important use for engines is in backup power systems. Reciprocating engines are simple stand-alone generating systems that can be easily and quickly started and this makes them popular for both high-security backup systems where energy must be available at all times and as less critical commercial backup systems where their low price makes them attractive. More recently these backup generators have increasingly been used for peak power too, providing the owner with power during periods when electricity demand peaks and the cost of electricity rises. This may be coupled with grid demand-side management where some consumers agree to allow their grid supply to be interrupted during high peak periods in return for a lower cost of electricity. An engine-based generating system can bridge the gap in supply.

On the other side of this equation, reciprocating engines – usually burning natural gas – are finding increasing application for grid power where they are used to support renewable generation. These engines can provide relatively clean power and have the ability to increase or decrease output quickly in response to the amount of wind or solar energy available.

A final, important market for reciprocating engines is in combined heat and power plants. These are usually small-scale installations in which the engine provides electric power while waste heat is captured and used for space heating or to provide hot water. However there are also large applications in which the heat energy captured is used for industrial processes.

In order to quantify the size of this market and provide some indication of the number and overall installed capacity of reciprocating

Table 1.1 Annual Reciprocating Engine Installed Capacity by Type	
Engine Type	Average Annual Installed Capacity, 2013−16 (MW)
Diesel engines	45,000−50,000
Dual fuel engines	100−500
Fuel oil engines	500−1000
Natural gas engines	4000−5000
Total	49,600−56,500
Source: *Decentralized Energy*[2]	

engines, globally, Table 1.1 presents figures for the annual capacity of reciprocating engines installed across the world, by engine type during the years 2013−16. Based on the figures in the table, the largest category in terms of capacity during this period was diesel engines with an annual installed capacity of between 45 GW and 50 GW each year. Diesel engines were popular throughout the world during this period, and earlier, but the biggest market for these engines in the middle of the second decade of the 21st century was in regions of Africa and Asia where power demand has been rising but the grid is weak, so supply is either unavailable or unreliable. In the developed world, emission regulations are limiting the use of diesel engines and cleaner engine types are preferred.

Natural gas is the clean fuel of choice for reciprocating engines but where the supply of this is intermittent or unreliable, dual fuel engines that can burn both liquid fuel and gas are popular. Markets for these engines include Africa and Asia and well as the Middle East and South America. The average annual installation of this type of engine was 100−500 MW between 2013 and 2016. Another category of reciprocating engine burns fuel oil. This is a heavy oil fuel and it is most widely used in countries that have their own oil supplies such as the Middle East. The latter remains the region where this type of engine is most popular but there are declining markets in both Asia and South America. Between 2013 and 2016 the average annual installed capacity for this type of engine was 500−1000 MW.

[2]Dina Darshini, How big is the gas-based distributed power generation market? And will it grow? Decentralized Energy, 30 January 2017.

Table 1.2 Breakdown of Annual Installation, 2013–16, by Engine Size

Engine Size	Number of Units	Annual Installed Capacity (MW)	Principal Fuel
50–500 kW	>10,000	10,000–30,000	Diesel
500 kW to 5 MW	>10,000	30,000–50,000	Diesel
5–10 MW	>100	1000–5000	Natural gas
10–50 MW	>100	1000–5000	Heavy fuel oil, dual fuel

Source: *Decentralized Energy*[3]

The other important category of engines is called natural gas engines, or simply gas engines, because they are designed to burn natural gas alone. Gas engines are cleaner than diesel engines and can be highly efficient. They tend to be more expensive too. These engines are popular in North America, Europe and Japan. Many of the plants where these engines are in use are combined heat and power stations but there is increasing demand for power-only plants to support grid renewable generation. Between 2013 and 2016 the average annual installed capacity for this type of engine was 4000–5000 MW, making them the second most popular category after diesel engines.

Complementing these figures, estimates by Navigant Research[4] suggest that the annual installation of diesel generating sets will increase from 63 GW in 2015 to 104 GW in 2024. As already noted, this is likely to be driven by nations where power demand is rising but supply is unreliable. The research company has also estimated that annual reciprocating gas engine installations will reach around 27 GW by 2024, a significant increase compared to the figures in Table 1.1.[5]

Further analysis of the annual installation of reciprocating engines is provided by the figures in Table 1.2. This shows annual installations for the same period, 2013–16, as in Table 1.1 but in this case broken down by engine size. For the smallest engines with generating capacities in the range 50–500 kW, the annual installation rate was over 10,000 units and the annual installed capacity was up to 30,000 MW.

[3]Dina Darshini, How big is the gas-based distributed power generation market? And will it grow? Decentralized Energy 30 January 2017.
[4]https://www.navigantresearch.com/research/diesel-generator-sets
[5]https://www.navigantresearch.com/research/natural-gas-generator-sets

Most of these engines are diesel engines. For engines in the next range, 500 kW to 5 MW, annual installations were also over 10,000 – in the report they were put at in the mid-10,000s whereas for the smaller size range the numbers were in the low 10,000s – while the additional annual capacity was between 30,000 MW and 50,000 MW. Again most of these engines burn diesel fuel.

The next size category, 5–10 MW, has many fewer engines installed each year with numbers somewhere in the low 100s. The majority of these engines were natural gas engines and annual additional capacity was between 1000 MW and 5000 MW. The largest engines, with generating capacities of between 10 MW and 50 MW, generally burn heavy fuel oil although some have dual fuel capability. More than 100 of these were installed each year in the period covered by the table and annual capacity additions were 1000–5000 MW.

While diesel engines continue to dominate for power generation there is a shift towards the clearer natural gas engines. This change is likely to become accelerated where natural gas is readily available, as the demand for cleaner power generation becomes more insistent. However there are still many regions where natural gas is not available and the only fuel option is diesel.

CHAPTER 2

Fuels and Energy Resources
for Reciprocating Engines

Reciprocating engines are heat engines and they require a source of heat energy if they are to be able to generate useful work. For an internal combustion engine this is normally supplied by a liquid or gaseous fuel that can be ignited in air within the combustion chamber, the cylinder of the engine, increasing the temperature and pressure of the gas inside the chamber and forcing the piston to move so as to expand the volume within the cylinder. External combustion engines often use the same fuels but they can also exploit a variety of other energy sources including solid fuels and solar radiation.

Many engines use liquid fuels such as gasoline or diesel because these are easily transportable and have a high energy density. This has made them attractive as automotive fuels both because a small reservoir of fuel can be used to carry a vehicle a long way and because the nature of the fuel makes it easy to distribute through gasoline stations along highways. In remote regions, such fuels can be supplied in bulk in large drums or tankers and this made the diesel engine the most important means of generating electric power for remote communities during the twentieth century. Another liquid fuel is liquefied propane gas (LPG). This is used in vehicles but is less likely to be used in stationary applications. Its advantage is that it produces lower emissions that either diesel or gasoline and can lead to longer engine life and lower maintenance costs.

The main gaseous fuel used in reciprocating engines is natural gas although there are other types of gas, in particular biogas, that can be used to fire them too. Although natural gas can be supplied in liquefied form as liquefied natural gas, LNG, it is normally provided via pipeline. Where a pipeline supply is available, gas engines are in common use. However there are many regions without any natural gas infrastructure and such regions cannot exploit these engines. For some of these, the use of bulk LNG supply is possible but in most cases and alternative liquid fuel, usually diesel, is preferred.

Piston Engine-Based Power Plants. DOI: https://doi.org/10.1016/B978-0-12-812904-3.00002-1

There are a number of situations in which a specialised source of a fuel gas encourages the use of a gas engine. The most obvious of these are landfill waste sites which produce a methane-rich gas by a process of anaerobic fermentation. This gas can be collected and used to generate power with a gas engine. Some waste processing plants also carry out the anaerobic fermentation of biomass waste in order to generate a gas for burning for heat and power production.

The balance between the use of liquid fuel and gaseous fuel for stationary power generation is also affected by fuel cost. This is particularly noticeable in a region such as the USA where the availability of cheaper natural gas as a consequence of shale gas recovery in the twenty-first century has promoted the use of gas engines for power generation. At the same time, as already noted, environmental concerns are reducing the use of diesel fuel.

External combustion engines can exploit almost any heat source so long as it can provide a sufficiently high temperature. For power generation, the most widely used external combustion engines are Stirling engines which can convert solar heat energy into electricity.

GASOLINE (PETROL)

Gasoline in the USA, called petrol in many other parts of the world, is the main liquid fuel used in spark ignition engines, one of the two main types of piston engines. The fuel is produced during the refining of crude oil. The amount of gasoline derived from the oil will depend on the source and the way the oil is processed. For a typical barrel of US oil, a 159 litres barrel can deliver up to 72 litres of gasoline. The latter is a mixture of hydrocarbons with between four and twelve carbon atoms in each molecule. These include alkanes, alkenes and cycloalkanes.

An important characteristic of gasoline is its octane rating. This is a measure of the ability of the fuel to resist spontaneous ignition in the chamber of the engine before it has been ignited by the spark plug. The smooth operation of the engine depends on the ignition taking place in a controlled manner. Spontaneous ignition, often called 'knocking', impedes this and if it takes place repeatedly it can damage an engine.

Knocking can occurs when the gas in the cylinder is compressed by the piston during part of the engine cycle. When a gas is compressed it heats up. How much it heats up — how hot it gets — will depend upon the compression ratio; the more highly the gas is compressed the hotter it will get. In a spark ignition engine this gas is a mixture of air and fuel and if the temperature becomes too high, the fuel may ignite spontaneously. Thus the composition of the fuel must be tailored to the engine design and compression ratio.

The octane rating of gasoline depends on its composition. Simple, straight chain saturated hydrocarbons (alkanes) have the lowest resistance to knocking and ignite the most readily. More complex branched hydrocarbons and aromatic hydrocarbons (these contain complex carbon ring structures) have higher resistance and therefore a higher octane rating. Some of the processes used during the refining of oil can convert the simpler hydrocarbons into more complex molecules. Gasoline is usually produced by blending refinery products from different treatment processes together to give a fuel with the desired properties.

The initial process carried out during the refining of oil is fractional distillation. This involves heating the crude oil slowly and allowing components to evaporate. The most volatile components tend to evaporate first and these are collected, followed by later fractions that have relatively higher boiling points. Straight run gasoline or naphtha, is the first product of direct distillation of crude oil. During the fractional distillation it has a starting boiling point of around 35°C and a finishing boiling point of 200°C. The naphtha collected between these boiling points has a relatively low octane rating. Its octane rating can be increased by using additives. The main historical additive used to increase octane rating is tetraethyllead. The use of this started during the 1920s but it was phased out towards the end of the century after the environmental and health damage caused by the lead was identified. Modern gasolines instead achieve a suitable octane rating by blending hydrocarbons with different properties.

In order to obtain these other hydrocarbons, naphtha is reformed. Reforming is usually performed using a catalyst, hence its common name of catalytic reforming. The process converts the straight chain hydrocarbons in the naphtha into branched chain and cyclic hydrocarbons.

Dehydrogenation may also be carried out in order to create unsaturated hydrocarbons. All these products have a higher octane rating than naphtha and can be blended with it.

Another common process is cracking of the naphtha. This can be carried out at high temperature and pressure without a catalyst or at lower pressure and temperature with a catalyst. Cracking generally breaks larger hydrocarbons into smaller ones but depending upon how it is carried out it can also increase their complexity and may dehydrogenate too.

The octane rating of a fuel is measured in a test engine. The hydrocarbon iso-octane, which has a branched structure, is arbitrarily assigned an octane rating of 100 (hence the name of the rating). N-octane, a straight chain hydrocarbon has a rating of 0. The octane rating of an unknown gasoline is then measured by comparing its performance in the test engine to various mixtures of the two standards. For example, if it performs like a mixture containing 10% n-octane and 90% iso-octane then it has an octane rating of 90.

An engine is usually specified so that it uses the fuel with a minimum octane rating consistent with smooth operation. The typical octane rating for modern road cars is 95 although some high performance vehicles use fuel with a rating of 98. There is no advantage in running a car with fuel of higher octane rating than it requires.

Another important property of gasoline is its vapour pressure. This should be high enough to allow the engine to start and operate at low temperatures but not too high that it causes vapour locks in the fuel feed system. Where seasonal temperature variations are wide, gasoline with different properties may be supplied at different seasons. Gasoline can deteriorate if stored for a long period as a result of oxidation. To counter this, fuel stabilizers may be added.

In some regions gasoline or petrol is mixed with a small amount of ethanol which is a cleaner, bio-derived fuel. Regulations mandate this in Brazil where around 25% ethanol is blended with gasoline. Ethanol is also widely used in the USA where a blend containing around 10% is common. Some US states require fuel to contain ethanol. In Europe the use of ethanol is most common in Sweden, Germany, France and Spain.

The energy content of gasoline is around 47 MJ/kg, based on the lower heating value. However the actual content will vary depending upon supplier and the season.

DIESEL

The second important branch of piston engines is the diesel engine. Diesel engines operate in a slightly different way to spark ignition engines. In a diesel engine, air alone is admitted into the cylinder before compression and it is compressed more highly than in a spark ignition engine, elevating the temperature of the air more. Fuel is only admitted at the end of the compression stroke when the air is at its hottest, hot enough so that the charge of fuel ignites spontaneously without the need for a spark. Since there is no fuel present when the air in the cylinder is compressed, knocking is not a problem. However the ignition properties of the fuel are still important. Diesel engines use a heavier fuel that spark ignition engines, and this fuel is commonly referred to as diesel.

There are a variety of diesel fuels in use today. The principal. historical diesel was, and is made by fractional distillation of crude oil, just like gasoline. However there are alternatives including synthetic diesel and biodiesel. Synthetic diesel can be manufactured from carbon based materials including coal, natural gas and biomass. Biodiesel is obtained from vegetable oil (sometimes animal fat too) that is converted into a liquid fuel suitable for use in diesel engines.

Petroleum diesel is the fraction that is condensed from the fractional distillation of crude oil after gasoline. It is produced at a fractional distillation temperature between 200°C and 350°C and contains carbon molecules containing, typically, between 8 and 21 carbon atoms. This makes the fuel relatively heavier than gasoline and less volatile.

Ignition takes place in a diesel engine when droplets of fuel are sprayed into the hot compressed air in the cylinder. How these droplets burn determines the efficiency and cleanliness of the engine. Different diesel fuels behave in different ways and this can be characterised, in the case of diesel fuel, by a rating called the cetane number which indicates its combustion quality. In this case the primary consideration is the delay in ignition of the fuel after it has been injected into the combustion chamber (usually but not always the cylinder). A higher cetane number equates to a more rapid ignition, a shorter ignition time and more complete combustion of the fuel. Shorter ignition time and complete combustion lead to more efficient and cleaner engines. The cetane number is usually derived from tests involving the fuel density and boiling or evaporation points.

Regular diesel fuel has a typical cetane rating of 48 and premium diesel 55. Biodiesel, depending upon the blend has a rating of 50 to 55 and the typical rating of synthetic diesel is 55. As with the octane rating, there is no advantage to running an engine using a fuel of higher cetane rating than it requires. Most modern diesel engines for road use operate with fuel of cetane rating between 45 and 55.

In addition to the cetane rating, diesel fuels are divided into three grades. Diesel No. 1 is a relatively volatile fuel containing molecules with between 8 and 19 carbon atoms. This is typically used in smaller, high speed diesel engines that operate at varying speeds and loads such as those for buses. Diesel No. 2 has a lower volatility, with molecules containing 9 to 21 carbon atoms. It is also used in high speed diesel engines but is usually restricted to those with a relatively constant speed and load. Diesel No. 4 is the heaviest diesel, often made by blending heavy distillate diesel and residual fuel oils. Typically it contains hydrocarbons with more than 25 carbon atoms. This type of fuel is used in medium and slow speed diesel engines that operate at constant load such as stationary power generation as well as ships and locomotives.

LIQUEFIED PETROLEUM GAS

Liquefied petroleum gas (LPG) is a third fuel that can be exploited to fire piston engines. It is used in spark ignition engines. The gas, normally supplied in liquid form under pressure, is a mixture of two hydrocarbons, propane (C_3H_8) and butane (C_4H_{10}). It may also contain iso-butane which has the same chemical composition as butane but a different structure. LPG is attractive because it has relatively lower carbon dioxide emissions than either gasoline or diesel. Its octane rating is between 90 and 110 depending on the proportion of the different hydrocarbons.

LPG is almost exclusively derived from fossil fuel sources. Its components can be isolated during the refining of crude oil and they are also present in natural gas as it emerges from the ground. The liquid has a slightly higher energy content than gasoline but its energy density is lower. Since its boiling point is below room temperature it must be pressurized to provide a liquid.

When it is used as an engine fuel, LPG is often called autogas. It is sold in many countries across the globe but the five largest markets are Turkey, South Korea, Poland, Italy and Australia. Between them, these five account for around half of the vehicles using this fuel. LPG is the third most popular fuel after gasoline and diesel but only accounts for around 3% of the global market for vehicles. It is used for stationary power generation too, but again the use is limited.

NATURAL GAS

Natural gas is an important fuel for stationary power generation and is used widely where gas pipeline systems are available. The gas, as it emerges from the ground, contains a mixture of smaller and larger chain hydrocarbons. The primary component is usually methane and this will provide up to 90% of pipeline gas. Other components include ethane, propane and butane as well as carbon dioxide. Reciprocating engines that burn natural gas are spark ignition engines and they are usually modified to optimize them for this fuel. The gas can also be used in road vehicles but there are only limited numbers of these operating across the globe. Natural gas can also be pressurized and liquefied.

There are other sources of methane rich gas in addition to natural gas. Landfill waste sites produce methane by a process of anaerobic digestion and this gas can be collected and used to provide heat and power using a gas engine. Animal and plant wastes can also be processed in digesters to provide a similar gas. This gas is usually referred to as biogas.

ENERGY SOURCES FOR EXTERNAL COMBUSTION ENGINES

External combustion engines, in which the heat to drive the engine cycle is provided from outside the engine, can generate energy from a variety of sources. The main engine of this type for power generation use is the Stirling engine. Stirling engines have been widely used in solar power generation using heat collected using large solar dish reflectors. They have also been developed for use in domestic combined heat and power systems where natural gas is used to generate heat that drives a Stirling heat engine with additional heat from the combustion process utilised for hot water and space heating. In principle the engines can exploit heat energy from any source but the applications where they offer a cost effective solution are limited.

CHAPTER 3

Types of Reciprocating Engine

Reciprocating engines come in many varieties but they share one common feature, power is produced through a piston moving backwards and forwards (reciprocating) inside a cylinder. That power is generated by pressure inside the cylinder and the pressure is normally produced by the combustion of fuel in air within the cylinder, causing the gas in the cylinder to heat up and expand. The way in which this combustion is initiated distinguishes spark ignition from diesel engines, the two principal varieties of reciprocating engine. There is another type of engine in which the expansion within the cylinder is generated using heat from outside the device. The most important of these 'external' combustion engines is the Stirling engine. Another variant, the steam engine, where the hot steam is generated outside the cylinder and admitted into it through a valve to create the pressure that drives the engine cycle, will not be considered here.

A piston moving within a cylinder produces a linear power stroke as the gas within the cylinder expands. How that linear motion is harnessed to provide usable power is key to the operation of an engine of this type. Moreover, the motion of the piston as a result of this expansion must at some stage be halted and reversed in order that another cycle can take place. How that stroke is controlled and how the piston is returned are features which further differentiate engine types.

The commonest type of piston engine is a rotating crankshaft engine in which the backwards and forwards motion of the piston is converted, using levers and bearings, into rotary motion in a shaft. The mechanical design ensures that the piston must return when it reaches the end of its power stroke, provided the engine continues to turn. These engines will usually have a large flywheel attached to one end of the shaft. This provides rotational inertia that ensures shaft rotation continues through from one power stroke to another.

Piston Engine-Based Power Plants. DOI: https://doi.org/10.1016/B978-0-12-812904-3.00003-3

There is another branch of piston engines called free piston engines in which the backwards and forwards motion is used directly rather than being converted into rotary motion. The power from these engines is often extracted via the exhaust pressure driving a turbine or with some form of hydraulic drive. These engines often use two opposing pistons and cylinders, one driving the piston in one direction, then the second driving it back. In this case the power strokes in the opposing cylinders must alternate.

For the more conventional rotating crankshaft engines, there are different engine cycles. The two most common are the two-stroke cycle in which each complete cycle of the engine consists of one in-stroke and one out-stroke, and the four-stroke cycle in which a complete engine cycle comprises two in-strokes and two out-strokes. There are also a number of six-stroke engine designs.

Another important variable is the number of cylinders in the engine. Some simple engines have a single cylinder. However the nature of the power generation in a crankshaft engine makes the power delivery from a single cylinder engine very uneven. More complex and sophisticated engines will have multiple cylinders with individual cylinders delivering power at different points during the rotation of the crankshaft in order to smooth the operation.

All reciprocating engines are heat engines that can be analysed in terms of the thermodynamic cycles which describe how heat can be used to provide power and work. The amount of power that can be extracted from a heat engine of this type depends in part on the temperature and pressure of the working fluid − in this case the gases in the cylinder of the engine − so controlling both the temperature and pressure during the engine cycle plays an important role in determining efficiency. For modern engines, emissions are also a key consideration. These will also be affected by the engine operating conditions but often in a contrary way to efficiency. For example, higher temperatures in the cylinder can provide higher efficiency but will lead to greater emissions of certain types.

INTERNAL COMBUSTION ENGINE FUNDAMENTALS

The simplest reciprocating engine comprises a single cylindrical chamber housing a piston. The engine cylinder is sealed at one end (but with valves to allow gases in and out) and open at the other end. A cylindrical-shaped disk of metal, the piston, is designed to fit closely within the cylinder to seal the open end and this piston is designed to move backwards and forwards easily within the cylinder. This it does in response to the pressure changes in the gas contained within the cylinder at various stages of the engine cycle. The outside of the piston is connected via levers to a shaft called the crankshaft through which rotary power is delivered. A cross section of a single piston engine is shown in Fig. 3.1. This is based on a spark ignition engine.

The operation of an engine of this type depends on repeating a cycle of events. These are controlled mechanically. For the commonest engine of all, the four-stroke spark ignition engine (the cycle is slightly different for a diesel engine) the cycle starts with a mixture of air and fuel being admitted into the cylinder while the piston moves (out) to increase the volume of the cavity within the cylinder. Once all the fuel and air has been drawn in, the piston starts to return, reducing the size of the cavity and compressing the gas and fuel mixture. When the piston reaches the top of its stroke and the volume of the cavity is at a minimum, an electrically generated spark ignites the fuel in the mixture, creating a controlled explosion that forces the piston to retreat.

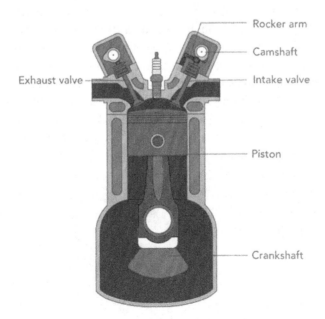

Figure 3.1 Cross section of a spark ignition piston engine cylinder and crankshaft. Source: US Office of Energy Efficiency & Renewable Energy.

This is the power stroke the drives the engine. The movement of the piston away from the cylinder head is eventually arrested mechanically and the piston begins to return once more to the top of the cylinder. As it does the combustion gases are expelled from the cylinder. As the piston reaches the top again, and all the gases have been expelled, the cycle repeats with a fresh charge of fuel and air being admitted.

In order for gases to be admitted and removed from the cylinder during this cycle there are valves fitted to the upper part of the cylinder chamber. These are controlled mechanically via a shaft (the camshaft) and levers (the rocker arms) that synchronise their movements to the movement of the piston within the cylinder. One valve, or a set of valves, is used to admit fuel and air into the cylinder while another valve or set allows these same gases to be expelled once combustion has taken place.

The key component of a reciprocating engine of this type, other than the cylinder and piston, is the crankshaft. This is a mechanical device that can convert reciprocating motion, backwards and forwards, into rotary motion or vice versa. The crankshaft has one or more arms (one arm for each piston) that stretch out perpendicular to the axis of the shaft, as shown in the model in Fig. 3.2. A rod (the connecting rod or conrod) is attached to the outside or bottom of each piston through a bearing while the other end of this rod is attached to the arm of the crankshaft through a second bearing. A schematic of this linkage is shown in Fig. 3.3. Operating through these two bearings, the linear motion of the piston is converted into rotary motion. The power stroke drives the piston away from the top of the cylinder, turning the crankshaft in the process. The

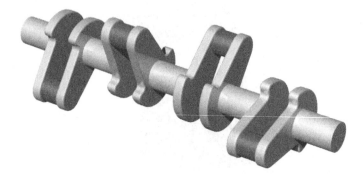

Figure 3.2 A three-dimensional model of a piston engine crankshaft. Source: Wikimedia.

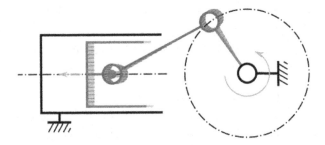

Figure 3.3 Schematic of the piston–crankshaft linkage. Source: Wikimedia.

mechanical coupling of the piston to the crankshaft then drive the piston back into the cylinder at the end of the power stroke, so long as the shaft continues to rotate. The rotation is maintained, and smoothed, by the fly-wheel attached to one end of the shaft.

The actual energy source for this type of engine is the fuel which burns in air, releasing large amounts of energy as heat. In the case of the spark ignition engine the fuel is mixed with air using a special mixing and metering system and a measured amount is introduced into each cylinder of the engine at the appropriate point in the engine cycle. A spark is then used to start a controlled explosion in which the chemical energy in the fuel is converted first into heat energy and then as the gases expand and act on the piston, into mechanical energy. This controlled explosion gives the engines its common name, the internal combustion engine.

As a result of the way in which the mechanical energy is fed to the crankshaft through levers, the rotational force on the shaft from each piston is extremely uneven. Most advanced engines have several pistons, synchronised mechanically to provide power sequentially as the shaft rotates. This helps to smooth the power delivery.

There are two principle types of internal combustion engine, the spark ignition engine and the diesel or compression ignition engine, each defined by the way in which fuel is admitted into the engine cylinder and how ignition of the air–fuel mixture is initiated. The spark ignition engine, as described above, uses an electrical spark to ignite the fuel–air mixture. The diesel engine takes a different approach. Instead of a fuel–air mixture, only air is admitted into the cylinder and this air is compressed much more highly than in the spark ignition engine. The higher compression ratio makes the air much hotter.

When the air is fully compressed, fuel is injected into it and ignites spontaneously in the hot air. This makes the engine design relatively simpler and these engines are, because of the higher compression ratio, potentially more efficient.

Many of the reciprocating engines in use today are based on what is known as a four-stroke engine cycle. This is the cycle that has been described above and involves the piston moving in and out twice during the full cycle. In consequence, power is delivered to the shaft of the engine only once during two revolutions. The four-stroke cycle is relatively complex but also allows the most sophisticated engines to be built. There is a simpler alternative, the two-stroke cycle. In this cycle a power stroke takes place during each revolution of the shaft. This type of engine is often used where a small, cheap source of mechanical power is required. However the engines do have some advantages over their more complex relatives, particularly a higher power to weight ratio. The two-stroke cycle is also used in some very large engines used for power generation since it is capable of high efficiency and tolerating very poor fuels.

ENGINE CYCLES

The internal combustion engine is a thermodynamic heat engine and as such belongs to the same category as steam turbines and gas turbines. However the physical nature of the reciprocating engine is very different to that of the turbines. The reciprocating engine principle can be traced back to the 17th century but development of the modern engine belongs to the latter half of the 19th century. Nikolaus Otto is generally credited with building the first four-stroke internal combustion engine in 1876. In doing so he established the principle still in use today.

The Otto cycle engine employs a spark to ignite a mixture of air and − traditionally − gasoline[1] compressed by the piston within the engine cylinder. This spark ignition causes an explosive release of heat energy which increases the gas pressure in the cylinder, forcing the piston outwards as the gas tries to expand. This explosion is the source of power, its force on the piston turning the crankshaft to generate rotary motion.

[1]Otto's engine probably burnt powdered coal but gasoline soon became the preferred fuel.

The Otto cycle was modified by Rudolph Diesel in the 1890s. In his version, air is compressed in a cylinder by a piston to such a high pressure that its temperature rises above the ignition point of the fuel which is then introduced into the chamber and ignites spontaneously without the need for a spark. This represents a simplification of the Otto cycle but is not without its complications, particularly from an emissions perspective.

At around the same time as the Otto and Diesel engines were being developed the first two-stroke engine cycle was also proposed. Two men, German engineer, Karl Benz and an Englishman, Dugald Clark, independently developed two-stroke engines in 1878. This cycle represents another important branch of the reciprocating engine family.

Four-Stroke Engines

In a four-stroke engine each piston of the engine, and there can be a large number depending on the particular engine type and application, is equipped with at least two valves, one to admit air or an air–fuel mixture and a second to exhaust spent gases after ignition. The opening and closing of these valves is mechanically synchronised with the movement of the piston backwards and forwards.

The four-stroke cycle derives its name from the four identifiable movements of the piston in the chamber, two of expansion and two of compression, for each full power cycle. These have already been outlined above but since this is the most important type of reciprocating engine, they are described again in more detail here, with diagrams to illustrate the process (Fig. 3.4).

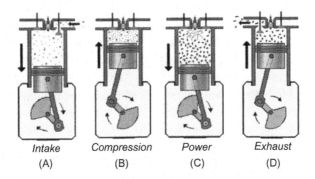

| Intake | Compression | Power | Exhaust |
| (A) | (B) | (C) | (D) |

Figure 3.4 The stages of a spark ignition engine cycle. Source: NASA.

Starting with the piston at the top of its chamber, and the chamber empty, the first stroke in an intake stroke in which either air (diesel cycle) or a fuel and air mixture (Otto cycle) is drawn into the piston chamber by movement of the piston to expand the volume of the enclosed space with the air or air–fuel mixture valve open. This valve closes at the end of the first stroke. The second stroke is a compression stroke during which the gases in the cylinder are compressed by the piston returning towards the top of its chamber. In the case of the Otto cycle, a spark ignites the fuel–air mixture at the top of this second stroke, creating an explosive expansion of the compressed mixture which forces the piston down again. This is the power cycle. In the diesel cycle fuel is introduced through a separate nozzle close to the top of the compression stroke, igniting spontaneously in the hot gas with the same effect. After the power stoke, the fourth stroke is the exhaust stroke during which the exhaust gases are forced out of the piston chamber though the second valve, now open. This closes at the end of the fourth stroke and the cycle begins again. In both spark ignition and diesel engines a large flywheel attached to the crankshaft stores angular momentum generated by the power stroke and this provides sufficient momentum to carry the crankshaft and piston through the three other stokes required for each cycle.

The shaft of an engine that is fitted with a single piston and cylinder will receive a power impulse once every two rotations, leading to a relatively uneven transfer of power. However if the engine has multiple cylinders, the cycle of each can be staggered relative to the others so that they deliver their power sequentially, leading to a much more even rotational motion. For a four-stroke engine it is normal for four (or a multiple of four) pistons to be attached to the crankshaft, with one of each set of four timed to produce a power stroke while the other three move through different stages of their cycles. In this case the power strokes of the four pistons should take place at $180°$ intervals during the rotation of the shaft for optimum delivery. The introduction of fuel and air and the removal of the exhaust gases are then controlled by valves which are mechanically timed to coincide with the various stages of the cycle on each cylinder.

Two-Stroke Engines

The two-stroke engine is much simpler than the four stroke. This is both an advantage and a limitation. In a two-stroke engine, intake and exhaust strokes are not separate and there are no valves. Instead the piston acts to close off or open ports through which fuel enters and exhaust gases exit the engine.

The case surrounding the engine and crankshaft (the crankcase) is often used to deliver the fuel to cylinder during each cycle. During the power stroke, ignition takes place while the piston is at the top of the cylinder and the air and fuel mixture is fully compressed. While in this position, more fuel and air can enter the engine casing through a port in the side. As the piston begins the power stroke this port is closed by the piston which simultaneously begins to put the mixture in the casing under pressure. Part way through the power stroke two new ports are opened into the cylinder, one is the exhaust port and the other is a port to admit fuel and air. When the piston reaches the bottom of the cycle and begins to return the fuel−air mixture, now pressurised compared to the combustion gases, enter the cylinder and help force the exhaust through the exhaust gas port. The exhaust and fuel ports close as the piston rises within the cylinder and the cycle repeats. Two-stroke engines are simpler than four-stroke engines because they do not require valves as the ports are controlled by the piston. However, some engines do use a small reed valve to control the fuel−air mixture entering the engine, with the reed opening when the pressure inside the crank case or cylinder is low compared to the feed pressure. There are also variations in the way the exhaust gases are forced out of the engine (scavenging), depending upon the gas flows and the shape of the piston head. The engine offers a better power to weight ratio than a four-stroke engine because power is delivered once for every cycle of the engine. Lubricating oil for the engine is often mixed with the fuel, so orientation of the engine is not critical for its operation. In consequence these engines can operate at any angle, making them useful for small devices such as chain saws. The design of larger two-stroke engines is more sophisticated and lubrication is maintained from within the engine. Small versions of the engine tend to be spark ignition while larger versions, often designed for high efficiency, are diesel engines. Two-stroke engines, particularly the smaller versions, have a narrow operating speed band so they are less flexible than four-stroke engines.

Six-Stroke Engines
In addition to the two main types of engine cycle there are also six-stroke engine cycles. The aim of these is to increase efficiency or reduce complexity. In the single piston six-stroke engine cycle, the normal four

strokes of an Otto cycle engine are followed by a further two during which a second fluid (such as steam or water) is injected into the cylinder, capturing heat that has been wasted during the preceding four piston movements. Other, compound six-stroke engines have two complementary pistons with one carrying out two strokes while the second performs the other four. A further variant is a five-stroke engine. This is a modified Otto engine in which a pair of cylinders share an extra cylinder placed between them. The exhaust from each Otto cycle cylinder is fed into the shared cylinder where it expands, providing additional work. There are in effect six piston strokes for each cycle, four in an Otto cylinder and two more in the ancillary cylinder. However the designers consider it to be a five-stroke cycle because the exhaust stroke of each Otto cylinder is synchronised with the expansion stroke of the additional cylinder and this is taken to be a single stroke. These engines are not widely used.

FREE PISTON ENGINES

A free piston engine is a reciprocating engine that does not use a crankshaft to control the motion of the piston(s) or to extract power from the engine. Instead, when the fuel is burnt in the cylinder of the engine, forcing the gas to expand and the piston to retreat, the force on the piston must be balanced by some other force acting on it within the engine. There are various ways this can be achieved. The simplest is to build a second, sealed, bounce chamber on the opposite side of the piston to the combustion cylinder and fill this with air. As the piston retreats it will compress this air and eventually the pressure will be high enough to force the piston to return into the combustion cylinder.

Another design uses two cylinders and combustion chambers with their open ends facing one another, each with a piston but with the two pistons joined together by a single rod. The combustion chambers are designed to fire alternately. When the cylinder on the right fires it forces the pair of pistons to the left, compressing the mixture in the left-hand cylinder. This then fires and returns the piston to the right. There are various ways that this to-and-fro motion can be used to provide a power source, with both hydraulic and electric power generation possible.

There is yet another configuration for the free piston engine, the opposing piston free piston engine in which two pistons share the same combustion chamber. When combustion takes place, the two pistons are forced in opposite directions and they are then returned by bounce chambers. Operation of this type engine requires the two pistons to be synchronised mechanically so that it remains balanced. This type of engine has been used in gas generators. The symmetrical design makes the engine vibration free.

Free piston engines are potentially more efficient than conventional reciprocating engines but practical engines cannot yet compete. There may be applications for these engines in hybrid vehicles and in the power generation industry if successful implementations can be developed.

ROTARY ENGINES

There are a number of engine designs that have been called rotary engines. One, often called a radial engine, has conventional four-stroke cylinders but the cylinders and pistons are arranged radially around the crankshaft. These engines always have an odd number of cylinders driving the shaft.

A second rotary engine is essentially the same as the radial engine but in this case it is not the crankshaft that rotates but the cylinders and pistons and the crankcase. The engines were popular in aircraft during the early decades of the 20th century, with the aircraft propeller being bolted directly to the crankcase.

A final variant is the Wankel engine. This uses an eccentric, triangular rotary piston or rotor inside an oval chamber, as shown in Fig. 3.5. The rotor moves within the chamber in such a way that it creates three individual combustion chambers. As the rotor turns, the volume of each changes. Each individual chamber is filled with a fuel and air mixture, the chamber becomes compressed and then the fuel−air mixture is ignited, and the expansion of the hot gases turns the rotor to allow this chamber to expand again and expel the exhaust gases. The engine has been used in motor vehicles and in some other applications but has never been widely adopted.

ENGINE SIZE AND ENGINE SPEED

The speed at which a piston engine operates will usually depend on its size. In general small units operate at the highest shaft rotational speed

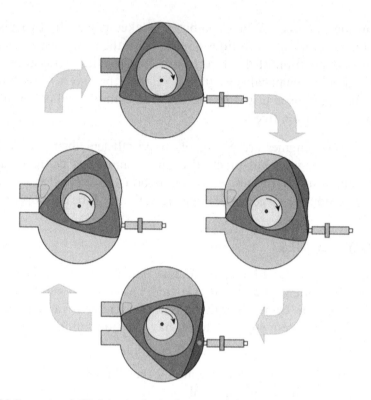

Figure 3.5 Cross-section of a Wankel engine showing the engine cycle. Source: Wikipedia.

and large units at the lowest shaft speed. In addition, in most situations a piston engine-based power unit will have to be synchronised to an electricity grid operating at 50 or 60 Hz so the engine speed will also be determined by one or the other of these rates. So for example a 50 Hz high-speed engine will typically operate at 1000, 1500 or 3000 rpm while the equivalent 60 Hz machine will operate at 1200, 1800 or 3600 rpm. These speeds allow the generators attached to the engines to synchronise with the grid operating frequency.

Engines for power generation are usually classified according to speed into one of three groups: high-speed, medium-speed and slow-speed engines. High-speed engines are the smallest and operate at up to 3600 rpm.[2] The largest slow-speed engines may run as slow as

---

[2]Reciprocating engines can reach much higher speeds. For example racing car engines can approach 20,000 rpm.

Table 3.1 Piston Engine Classification by Size and Speed		
	Engine Size	Engine Speed (rpm)
High speed	1 kW to 8.5 MW	1000–3600
Medium speed	1–35 MW	275–1000
Slow speed	2–65 MW	50–275
Source: *US Environmental Protection Agency.*		

50 rpm. Typical speed and power ranges for each type of engine are provided in Table 3.1.

Engine performance varies with speed. High-speed engines provide the greatest power output as a function of cylinder size, and hence the greatest power density. However the larger, slower engines are more efficient and last longer. Thus the choice of engine will depend very much on the application for which it is intended. Large, slow or medium-speed engines are generally more suited to base-load generation but it may be more cost-effective to employ high-speed engines for backup service where the engines will not be required to operate for many hours each year.

In addition to standby service or continuous output base-load operation, piston engine power plants are good at load following. Internal combustion engines operate well under part load conditions. For a gas-fired spark ignition engine, output at 50% load is roughly 8%–10% lower than at full load. The diesel engine performs even better, with output barely changing when load drops from 100% to 50%.

CHAPTER 4

Spark Ignition Engines

Spark ignition engines are the most popular reciprocating engines. Engines of this type can be found in many road cars as well as in some larger vehicles. They are used in small mechanically driven tools such as lawn mowers and chain saws and they can be used for a variety of power generation services including emergency backup and grid support as well as base-load supply.

The engines can burn a variety of fuels. The most common is gasoline (petrol) but they can burn a range of gaseous and liquefied gas fuels including natural gas, propane, biogas and landfill gas. Liquid fuel, gasoline, is the preferred fuel for most mobile applications because of its relatively high energy density but for stationary applications the use of gaseous fuel, particularly natural gas, is common and becoming increasingly popular.

Most spark ignition engines are four stroke; two-stroke engines of this type are generally only used for small mechanical devices. The latter are also popular in motor cycles although most advanced motor cycles now use four-stroke engines. For power generation applications, engine sizes range from less than 1 kW to around 10 MW. Reciprocating engines larger than this are normally large, slow-speed diesel engines.

The spark ignition engine premixes the fuel with air before admitting the mixture into the cylinder of the device. Various mechanical and electrical systems have been developed to carry out this mixing. The fuel mixture is then compressed before being ignited with an electrical spark. In some more complex engines there is a pre-ignition chamber where a fuel-rich mixture is ignited first, with this ignition then spreading into the main cylinder. The precise fuel and the composition of the air—fuel mixture will influence performance parameters such as the efficiency and the level and type of engine emissions.

Another important variable is the compression ratio which indicates the degree to which the gases are compressed before they are ignited.

Piston Engine-Based Power Plants. DOI: https://doi.org/10.1016/B978-0-12-812904-3.00004-5

Compressing a gas raises its temperature and the compression ratio will affect the propensity of the air–fuel mixture to ignite spontaneously and so must be carefully controlled. The likelihood of this can also change with engine load and speed.

Efficiency and power output have traditionally been the key considerations in engine design. Today emissions are also an important but contrary consideration; optimising an engine for low emissions can lead to lower efficiency or lower engine power, so finding a compromise that balances the different demands is a vital part of modern engine engineering.

SPARK IGNITION ENGINE FUNDAMENTALS

The spark ignition engine exploits the Otto cycle for a four-stroke engine. The cycle has been described in Chapter 3, Types of Reciprocating Engine but the various stages will be examined in greater detail here. The four stages or strokes of the cycle are shown again in Fig. 4.1. These comprise an intake stroke when a fuel–air mixture is drawn into the engine, a compression stroke when the mixture is compressed, a power stroke when the mixture is ignited and expands and an exhaust stroke when the combustion gases are expelled from the cylinder.

There are a number of commonly used technical terms associated with this cycle. The stroke of the engine is the distance the piston moves from the top of the cylinder to the bottom of the cylinder. This is twice

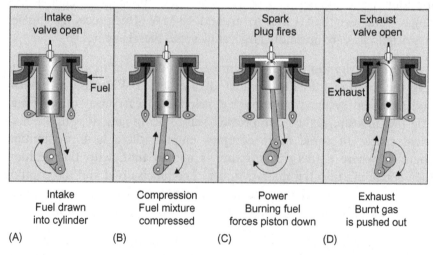

Figure 4.1 The Otto four-stroke engine cycle.

the distance between the centre of the crankshaft and the centre of the bearing attaching the connecting rod to the crankshaft. The position when the piston is at the top of its stroke is called top dead centre (TDC) and the position at the bottom of the stroke is called bottom dead centre (BDC). Since the force developed in the piston during the power stroke is applied to the shaft through the arm as a rotational moment, at both TDC and BDC there is no rotational moment about the crankshaft. On the other hand, the rotational moment will be greatest midway between TDC and BDC. The volume within the cylinder is at its minimum when the piston is at TDC. It is at its maximum when the piston is at BDC. The engine displacement, normally referred to as the size of the engine, is the difference in volume between TDC and BDC. (It is also the length of the stroke of the engine multiplied by the cross-sectional area of the piston which is defined by its diameter, or bore.) For a multi-cylinder engine, the volume of all the cylinders is added together to give the total engine size.

The spark ignition engine is a heat engine which means that is converts heat energy into a mechanical output, or in thermodynamic terms, into work. The actual engine cycle is complex to analyse but it can be simplified or idealised. A thermodynamic schematic of the ideal Otto cycle pressure–volume diagram is shown in Fig. 4.2. This can be used to determine the efficiency of the engine. The analysis essentially ignores the intake and exhaust strokes of the engine. These involve either drawing air/fuel into the cylinder or expelling the combustion products. In both cases, a

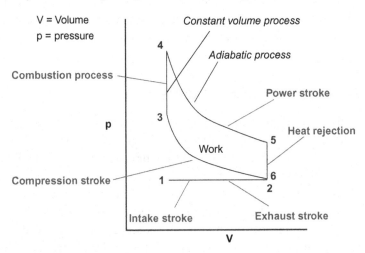

Figure 4.2 Ideal thermodynamic pressure–volume diagram for the Otto cycle. Source: NASA.

valve is open so and the gas in the cylinder is drawn in or expelled at constant pressure and it is assumed that during these stokes no energy is either produced or used. Only the volume within the cylinder changes. These strokes are represented between points 1 and 2 in Fig. 4.2.

The two remaining strokes define the engine. The first is a compression stroke, 2−3 in the diagram, in which the fuel−air mixture is compressed. This reduces its volume, increases its pressure and increases the temperature because the temperature of a gas rises when it is compressed. At the end of the compression stroke the fuel is ignited and the chemical reaction that takes place releases large amounts of heat. This is considered, in the idealised cycle, to take place instantaneously while the volume within the cylinder does not change. However the pressure of the gases rises dramatically (3−4). The hot, high-pressure gases then force the piston away, expanding the volume within the cylinder as the pressure drops (4−5). At the end of this stroke the exhaust valve is opened and any excess heat and pressure is released (5−6). This is considered in the ideal case to be another instantaneous process. Finally, the exhaust stroke takes place (2−1) while the combustion gases are expelled at constant pressure.

In this ideal form, there are two phases in which the both volume and pressure change. The compression stroke is the first and during this work is done on the gas to compress it and so energy is expanded. The second phase is the power stroke in which the expanding gases force the piston to move. This generates power. Mathematically, the net amount of useful work that the engine provides is that generated in the power stroke minus that used during the compression stroke. It is represented graphically by the area within the cycle diagram in Fig. 4.2.

FUEL PREPARATION AND COMBUSTION

In order for a spark ignition engine to operate smoothly, the composition of the air−fuel mixture must be carefully controlled and the rate at which it is delivered to the engine cylinders must be controlled too. In virtually all engines until the 1980s these processes were controlled mechanically using a device called a carburettor. Since then most engines have converted to fuel injection because this allows closer control of the fuel mixture and therefore closer control of the emissions from the engine. Modern engines use electronically controlled injection systems.

The composition of the fuel–air mixture in the cylinder may be close to the stoichiometric ratio required for complete combustion of the fuel in air but more often it will contain a significant excess of air. In common with all thermodynamic heat engines, the efficiency that a reciprocating engine can achieve increases with the temperature of the working fluid, in this case the combustion gases in the cylinder. For a spark ignition engine the highest cylinder temperature is reached when the air to fuel ratio is around 15:1, the ratio at which a stoichiometric amount of oxygen is available to react with all the carbon and hydrogen within the fuel. An engine which operates with this air to fuel mixture is described as a rich-burn engine. A rich mixture leads to the highest temperature, and the highest engine power, but also leads to the greatest formation of nitrogen oxides (NO_x) as well a significant amounts of carbon monoxide and unburnt hydrocarbon particles as a result incomplete combustion of some of the fuel. Under most circumstances therefore, engines operating continuously on a rich mixture will require emission control systems to limit release of these potential pollutants.

If engine emissions are to be reduced during combustion, then the combustion temperature must be lowered and a greater amount of oxygen introduced to ensure complete combustion of the fuel. Such engines are described as lean-burn engines and can operate with an air to fuel ratio of between 20:1 and 50:1 depending upon the fuel, significantly higher than in the rich-burn engine. The greater proportion of air lowers the overall combustion temperature (there will be relatively less fuel entering the combustion chamber in the lean mixture), reducing the production of nitrogen oxides from nitrogen in air and providing the conditions for much more complete combustion of the fuel. This will reduce the amounts of carbon monoxide and unburnt hydrocarbons in the exhaust gases. Against this, the lower temperature reduces overall efficiency and power. Lean-burn engines achieve a typical efficiency of only 28% (LHV),[1] compared to up to 42% (LHV) for a rich-burn engine. An engine tuned for maximum efficiency will produce roughly twice as much NO_x as one tuned for low emissions. Typical NO_x emission levels for spark ignition engines are 45–150 ppmV.

---

[1]The energy content of a fuel may be expressed as either the higher heating value (HHV) or the lower heating value (LHV). The HHV represents the energy released when the fuel is burned and all the products of the combustion process are then cooled to 25°C. This energy then includes the latent heat of vaporisation released when any water produced by combustion of, for example, natural gas is condensed to room temperature. The LHV does not include this latent heat and is hence around 10% lower than the HHV in the case of natural gas.

Until the 1980s the main device for mixing fuel with air before feeding it into the cylinders of the engine was called a carburettor. This was developed and patented by Karl Benz in 1888. It is a mechanical device that uses the Venturi effect to draw fuel into the air as shown in Fig. 4.3. Air for the engine flows through a tube in which there is a constriction called a Venturi. As it passes through this constriction, the velocity of gas increases but its pressure drops. A small jet in the side of the Venturi allows fuel to be drawn into the air flowing past it: the fuel is sucked from this small jet as a consequence of the pressure drop as air flows through the Venturi. The fuel is fed from a small float chamber which operates a little like a toilet cistern. Meanwhile there are two valves in the air-flow tube. One before the Venturi is called a choke because closing it chokes-off the air flow. This leads to relatively more fuel mixing with the air than if it were fully open, leading to a richer fuel−air mixture. The second valve, after the Venturi, is

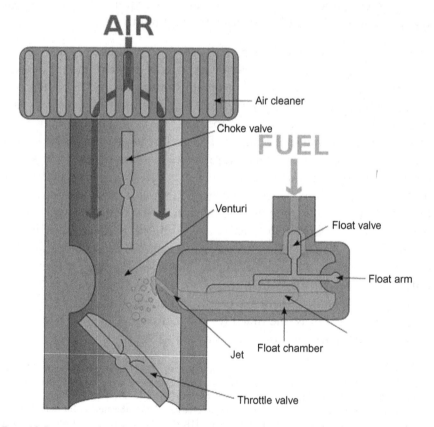

Figure 4.3 Cross section of a carburettor. Source: Wikimedia.

called the throttle. This controls the absolute amount of air that can enter the engine, whatever the fuel−air ratio. If more fuel and air are allowed in, the engine runs faster or accelerates.

An important feature of the spark ignition engine is that its speed is self-limiting because the speed or power of the engine is determined by the amount of fuel that is fed to it. Since the engine is fed with a mixture of fuel and air, controlling the quantity using the throttle will control the engine speed.

More modern engines use an electronically controlled fuel injection system to ensure that the proportion of fuel in the air is exactly correct during each moment of the engine's operation. This enables tight control of emissions. For a conventional automobile engine the most power is obtained with a relatively rich fuel mixture while the best economy is obtained with a leaner mixture. However the efficiency will also depend upon how well the fuel and air are mixed. The better the mixing, the faster and more completely combustion of the fuel can take place. The amount of time available for this to occur is very short. For engine running at 4000 rpm, there will be 66 revolutions and 33 firings of each cylinder each second, allowing 8 ms for each power stoke. The actual combustion probably takes place in less than one-tenth of this time.

Combustion of the mixture in the cylinder is normally carried out using a spark plug. This has two electrodes which are inside the combustion chamber at the top of the cylinder. At a pre-determined point during each cycle a spark is generated across these electrodes to ignite the mixture. The combustion must then spread almost instantaneously through the compressed fuel−air mixture. More technically complex engines can use a pre-ignition chamber in which a small amount of an air−fuel mixture rich in fuel is admitted and ignited. This pre-ignition then spreads into the main cylinder where an air−fuel mixture containing a much greater proportion of air is ignited.

The point at which ignition takes place, the 'timing' of the engine, is important for a number of reasons. Ideally it should take place when the piston is at TDC but this may not allow sufficient time for combustion to take place fully. It is common, therefore to 'advance' the timing so that ignition takes place slightly before the piston reaches TDC. Timing will also be affected by another important engine parameter, the compression ratio.

The compression ratio of an engine is the ratio of the volume of the cylinder at BDC divided by the volume at TDC, or the amount by which the fuel−air mixture has been compressed. Higher compression ratios lead to greater power from the engine but also lead to a higher gas temperature in the cylinder before ignition and this can cause spontaneous ignition or knocking. The compression ratio of a spark ignition engine (the amount by which the air−fuel mixture is compressed within the cylinder) is normally limited to a maximum of between 9:1 and 12:1 to avoid knocking. Lean natural gas−air mixtures have a much higher resistance to knocking than stoichiometric mixtures and can therefore tolerate higher compression ratios than gasoline.

Timing can also be used to help alleviate knocking. If the timing is advanced, then engine ignition can take place before the mixture in the cylinder reaches the temperature at which it will knock. In older engines the timing was fixed mechanically but in modern engines this can also be controlled electronically.

NATURAL GAS-FIRED SPARK IGNITION ENGINES

The natural gas-fired spark ignition engine has become popular for distributed generation and for a range of other power generation duties in recent years. The primary reason for this is its low emission performance compared to both gasoline engines and diesel engines.

As a fuel for a spark ignition engine, natural gas has an octane rating of around 120−130 depending on the source and composition. This allows natural gas-fired engines to operate at higher compression ratios than conventional gasoline engines and compression ratios of between 12 and 15 are popular. A higher compression ratio allows higher power to be delivered from an engine of a fixed size and this can lead to higher efficiency.

In addition to the high compression ratio, natural gas engines can be operated with much leaner fuel to air mixtures than gasoline engines. The leaner mixture leads to a lower combustion temperature which reduces efficiency compared to a gasoline engine but this is compensated for by the higher compression ratio. More importantly, the lower temperature can lead to lower production of nitrogen oxides and the excess air and oxygen in the mixture means that it is easier to achieve complete combustion. This reduces the quantities of unburnt hydrocarbons and carbon monoxide, both of which are products of incomplete combustion.

Since natural gas is a gas, effective mixing with air is much easier to achieve than with a high vapour pressure liquid fuel such as gasoline. This allows for smoother combustion. However problems will arise if the ratio of air to natural gas becomes too high because then the mixture may not ignite, or may only ignites partially, leaving unburnt gas in the exhaust.

DUAL FUEL ENGINES

Diesel engines can operate with much higher compression ratios than spark ignition engines and this allows them to achieve higher efficiencies. The large disparity in efficiency between a spark ignition engine and a diesel engine has prompted engine developers to search for a way of achieving the efficiency of a diesel engine in spark ignition engine. This is the origin of the dual-fuel engine which has been the most successful of these hybrids (Fig. 4.4).

A dual fuel engine is an engine designed to burn predominantly natural gas but with a small percentage of diesel as a pilot fuel to start ignition. The engines operate using a hybrid of the diesel and the Otto cycles. In operation, a natural gas–air mixture is admitted to the cylinder during the intake stroke, then compressed during the compression stroke. At the top of the compression stoke the pilot diesel fuel is admitted and ignites spontaneously, igniting the gas–air mixture to create the power expansion. Care has to be taken to avoid spontaneous ignition of the natural gas–air mixture, but with careful design the engine can operate at close to the compression conditions of a diesel engine, with a high-power output and high efficiency, yet with the emissions close to those of a gas-fired spark ignition engine. However efficiency tends to fall and emissions of unburnt hydrocarbons and carbon monoxide rise at part load.

Typical dual fuel engines operate with between 1% and 15% diesel fuel. Since a dual fuel engine must be equipped with diesel injectors, exactly as if it were a diesel engine, a dual fuel engine can also burn 100% diesel if necessary, though with the penalty of much higher emissions.

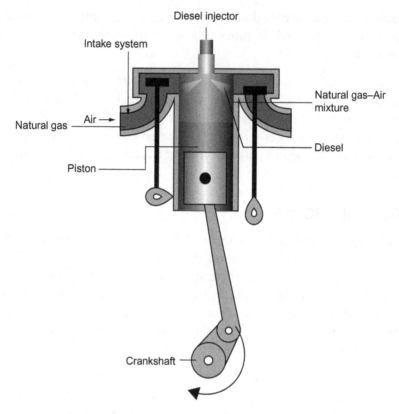

Figure 4.4 Schematic of a dual-fuel engine.

TWO-STROKE CYCLE CONSIDERATIONS

The moving components of all reciprocating engines must be kept lubricated to reduce friction between components and in the extreme to stop them binding together and failing. Four-stroke engines usually have a sophisticated oil circulation system that ensures all the major components including the pistons and cylinders are kept lubricated. Some of the oil that is used to lubricate the engine will inevitably be burnt and end up in the exhaust but in modern four-stroke engines the amount is normally limited.

Two-stroke engines are much simpler and the lubrication system is often simple too. In fact in many small two-stroke engines, the oil that is needed to keep the pistons and cylinders well lubricated is actually mixed with the fuel so that it is left behind when the combustion mixture burns. This system generally leads to much higher levels of burnt, or more often partly burnt, oil in the exhaust of the engine. For this reason small two-stroke engines have relatively high levels of certain emissions.

POWER APPLICATIONS FOR SPARK IGNITION ENGINES

Spark ignition engines can be used in a range of power generation applications. Small engines, based on automotive engines, are produced in large numbers and are relatively cheap to buy and reliable but are inefficient and have relatively short lives. These engines are most often used for standby and backup duties where the engine will supply power if the grid supply fails. Under these circumstances neither efficiency nor lifetime is likely to be of primary importance since the engines are not required to be in service either very often or for very long. System sizes can range from a few kW up to 100 kW though most are smaller than this.

For more mainstream power generation applications, much larger engines, in the 250 kW to 10 MW range are available, many designed to burn natural gas. These engines are designed to longer lives, lower maintenance costs and to achieve high efficiency. Consequently they are also much more expensive. The largest of these engines can be highly efficient, some offering up to 49% efficiency for conversion of natural gas into electricity. These engines can be used for distributed generation, for local combined heat and power and in some cases for grid support. Gas engines can also be adapted to burn landfill gas and biogas. Some can also be adapted to burn 'furnace gas', usually a mixture of hydrogen and carbon monoxide, produced during smelting of metals.

POWER APPLICATIONS FOR SPARK IGNITION ENGINES

CHAPTER 5

Diesel Engines

The diesel engine, sometimes called the compression ignition engine, was developed by Rudolph Diesel who built his first engine in 1892 and patented it in 1893. When Diesel developed his engine, current steam and gasoline reciprocating engines achieved around 10% efficiency. He believed that it should be possible to convert more of the energy released from fuel into power by exploiting thermodynamic heat engine principles developed by Carnot. His first engine was around 25% efficient. Modern diesel engines can achieve twice this efficiency.

The diesel engine is a reciprocating engine, just like a spark ignition engine, and it shares many of the same parts. Where it differs is in the manner of ignition of the fuel. This takes place spontaneously as a result of the high temperature reached by the air in the cylinder of the engine during compression. In consequence, no ignition system is required.

To achieve the high temperature necessary, the engine must compress the air in the cylinder very significantly. Diesel's original engine used coal dust as the fuel and had a compression ratio of roughly 100. This, combined with the absence of any cooling system, meant that this prototype engine was potentially extremely dangerous and it exploded, almost killing its inventor. He soon adapted it, reducing the compression ratio to 37, adding a cooling jacket and using a liquid fuel. This version proved a success.

In order to cope with the higher pressures inside the engine, a diesel engine is more robust, and hence heavier than a gasoline engine. This makes it more expensive, but this is offset by the higher efficiency. Another advantage is that the engine can burn heavier fuel than a gasoline engine and this heavier fraction, obtained from fractional distillation of petroleum after gasoline, has traditionally been much cheaper than the lighter gasoline. The price differential is much smaller in the 21st century.

Piston Engine-Based Power Plants. DOI: https://doi.org/10.1016/B978-0-12-812904-3.00005-7

Diesel engines have been used for heavy transportation applications such as locomotives and marine engines and many of these heavy engines have been adapted for power generation use too. Most are four-stroke engines but some of the very heavy engines are two-stroke engines.

DIESEL ENGINE FUNDAMENTALS

The diesel engine is a heat engine that utilises the properties of a gas to covert heat energy into mechanical energy. When a mass of air is contained in a restricted volume such as the cylinder of an engine and then heat is added to it, the pressure of the gas increases. This increase in pressure can be exploited to generate a mechanical force, power. The cross section of a diesel engine cylinder is shown in Fig. 5.1.

Most diesel engines have four strokes, exactly like the spark ignition engine. For an idealised engine these four strokes are an intake stroke when air is drawn in to the cylinder through a valve as the piston moves away from the top dead centre position (TDC – see Chapter 4) towards the bottom dead centre position (BDC). When it reaches BDC, the valve closes[1] and the piston returns towards TDC, compressing the air inside the cylinder as it does so. When it reaches TDC again, diesel fuel is injected into the compressed gas, which is now very hot as a result of being compressed and the fuel burns, increasing the temperature and hence the pressure inside the cylinder further. This additional pressure on the piston head forces the piston back towards the BDC position, providing the power stroke of the engine that can be harnessed to provide mechanical drive. Finally, at BDC the piston returns again, this time with a second, exhaust valve open when the air and combustion products are expelled from the cylinder.

The stages of the cycle can be represented by a pressure–volume diagram that represents the gases inside the engine cylinder. This is shown in idealised form in Fig. 5.2. This diagram ignores the first stroke of the cycle which draws air into the cylinder, and the last stroke which expels the combustion gases because these two strokes, ideally, involve no exchange of energy. (In practice, they do require energy to complete but the amount is small compared to the energy

---

[1]In real engines the opening and closing of the valves is actually offset from BDC and TDC.

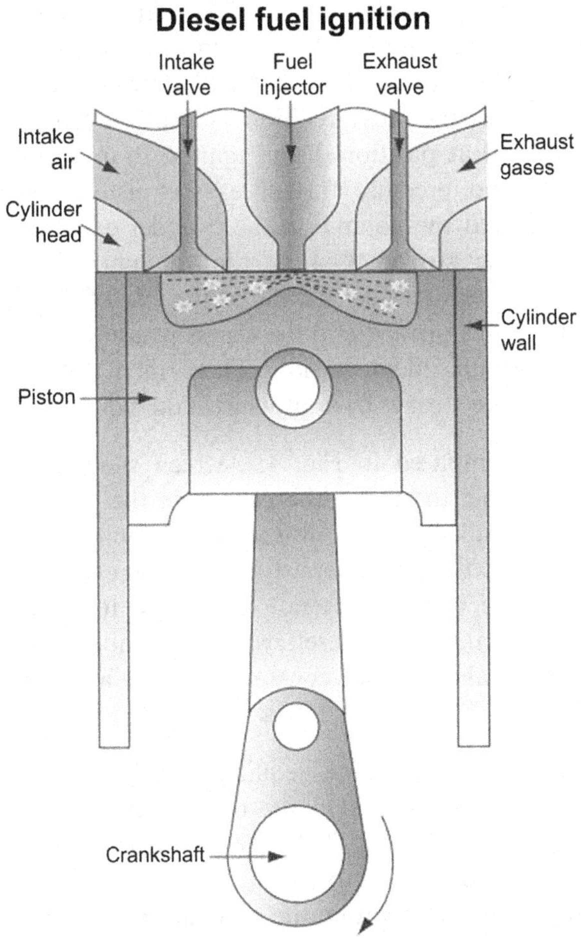

Diesel fuel ignition

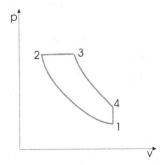

Figure 5.1 Cross section of a diesel engine cylinder. Source: US Department of Energy.

Figure 5.2 An idealised thermodynamic pressure–volume diagram for a diesel engine. Source: Wikimedia.

exchange involved in the other two strokes.) At position 1 in the diagram the cylinder is assumed full of air and this air is compressed by the piston as it moves to position 2. This compression stroke reduces the volume, increases the pressure and increases the temperature of the air. Fuel is injected at position 2 and ignites, dramatically increasing the temperature and pressure further as the piston begins to move away from TDC and the volume in the cylinder expands. This is followed by the power stroke, 3−4, as the volume inside the cylinder increases and the pressure falls. Finally at the end of the power stroke, 4, the exhaust valve is opened and the excess pressure is released, again instantaneously in this ideal version. There then follow the exhaust stroke and the intake stroke, both of which take place at position 1.

If Fig. 5.2 is compared to Fig. 4.2, which shows the cycle for a spark ignition engine, the only difference is in the change that occurs when combustion takes place. In the spark ignition engine this is assumed to take place instantaneously within the cylinder, at constant volume because the piston does not have time to move during the explosive combustion. In the diesel cycle the combustion takes longer and is assumed to take place at constant pressure as the piston moves away from TDC.

The compression stroke, 1−2, requires energy to be used to compress the gas in the cylinder. On the other hand, the power stroke, from 2 to 4, generates power. The net amount of power available for useful work is the difference between the two. This is represented mathematically by the area within the cycle in the diagram.

DIESEL ENGINE COMBUSTION AND TURBOCHARGING

The primary difference between a spark ignition engine and a diesel engine is the manner in which ignition occurs. The diesel engine does away with the ignition system and instead relies on the temperature of the air in the cylinder being high enough to ignite the fuel spontaneously. This fuel is added at the end of the compression stroke whereas in a spark ignition engine the cylinder already contains an air−fuel mixture.

In order to achieve ignition, the air in the cylinder is compressed much more than in a spark ignition engine and it is the compression that raises the temperature. The compression ratio of a diesel engine is

Table 5.1 Four-Stroke Engine Performance Parameters		
	Diesel Engine	**Spark Ignition Engine**
Typical size range	1 kW to 80 MW	1 kW to 6.5 MW
Efficiency	30%–48%	28%–42%
Compression ratio	14:1 to 25:1	8:1 to 12:1

typically between 14:1 and 25:1 (Table 5.1). In order to withstand the higher pressure, the engine components in a diesel engine must be stronger than in a spark ignition engine. This makes the engines heavier and more expensive than gasoline engines. However it also leads to higher efficiency so that a diesel engine can achieve 50% fuel to energy conversion efficiency, significantly higher than a gasoline engine.

Another ramification of the higher compression ratio is that the combustion of fuel leads to a higher temperature in the cylinders that would be found in a spark ignition engine. This means that a highly efficient cooling system is required.

Fuel is introduced into the cylinders of a diesel engine using injectors. These are precision components that must be able to accurately control the amount of fuel that is injected during each cycle of the engine. Control is important because it is the quantity of fuel that determines the speed at which the engine runs. Fuel within the fuel system also acts as a coolant for the fuel system and more fuel is generally pumped around the fuel system than is needed to supply the engine.

In order to gain the highest efficiency and to ensure that the high volumetric air requirements of the diesel engine can be met, many diesel engines are turbocharged or supercharged. A turbocharger is a pump that draws air into the engine and pressurises it. The pump is driven by an impeller that is placed in the exhaust gas stream of the engine. This essentially captures waste energy from the engine exhaust and uses it to boost the air input. The higher pressure air helps supply the cylinders with the air they need. Some diesel engines have an air input port part-way down the cylinder in addition a valve at the top of the cylinder. This is blocked by the piston during part of the cycle but opens as the piston falls towards BDC. Air that is swept into the cylinder through this port can help sweep out (scavenge) the combustion gases. Higher pressure intake air makes this more efficient.

A supercharger, the alternative to a turbocharger, also acts to pump air into the engine. However instead of being driven from the exhaust gases it takes its power directly from the engine crankshaft.

Compression of air with a turbocharger or supercharger will raise its temperature and this can affect the efficiency of the engine by lowering the density of the mixture entering the combustion chamber. In order to avoid this a cooler, often called the aftercooler or intercooler, is used to cool the air, which often exits the turbocharger at 200°C, to around 30°C before it enters the engine cylinders. The intercooler often has two stages, particularly if engine heat is being captured for cogeneration. This allows the first stage to provide heat at a relatively high temperature.

The temperature inside the cylinder of a diesel engine during ignition rises much higher than the temperature in a spark ignition engine. As a consequence the production of NO_x is much higher. Typical levels are 450–1800 ppmV or 10 times higher than for the equivalent spark ignition engine. Diesel engines therefore require extensive emission control systems if they are to comply with air quality regulations, particularly when the units are operating in an urban environment. Against this, diesel engines are capable of burning bio-diesel, a carbon dioxide emission neutral fuel. This can be attractive in some situations.

The practical efficiency of the diesel engine ranges from 30% (HHV[2]) for small engines to 48% (HHV) for the largest engines. Some very large two-stroke diesel engines can reach higher efficiencies still with the addition of a steam turbine bottoming cycle. Higher pressures and temperatures could conceivably produce yet higher efficiency but as with other types of thermodynamic engine, materials are the limiting factor. Very high combustion temperatures and pressures require exotic materials that are able to withstand them without deforming. These are much more expensive than traditional steels and increase the cost of the engine significantly, undercutting its economic advantage.

---

[2]The energy content of a fuel may be expressed as either the higher heating value (HHV) or the lower heating value (LHV). The HHV represents the energy released when the fuel is burned and all the products of the combustion process are then cooled to 25°C. This energy then includes the latent heat of vaporisation released when any water produced by combustion of, for example, natural gas is condensed to room temperature. The LHV does not include this latent heat and is hence around 10% lower than the HHV in the case of natural gas.

Diesel engines can be built to larger sizes than spark ignition engines, with high-speed machines available in sizes up to 4 MW and slow-speed diesels up to 80 MW. Large slow-speed engines can have enormous cylinders. For example, a nine-cylinder, 24 MW engine used in a power station in Macau has cylinders with a diameter of 800 mm.

Diesel engines can burn a range of diesel fuels including both oil-derived fuels and biofuels. Smaller, high-speed engines normally use high-quality distillate but the large slow-speed engines can burn very low-quality heavy fuel oils which require a much longer combustion time to burn completely. These fuels tend to be dirty and plants burning them usually require additional emission mitigation measures.

ENGINE TIMING AND SPEED CONTROL

The combustion of fuel in a diesel engine takes place after the fuel has been injected into the hot compressed air within the cylinder of the engine. However combustion will only take place once the fuel has vaporised. This takes time because vaporisation of the initial droplets cools the compressed gas around it. In order to take account of this and ensure that the combustion takes place at the correct position of the piston during the cycle, fuel injection begins before the piston has reached TDC[3] and finishes at TDC or marginally afterwards.

Once ignition of the fuel has started, the additional heat from the combustion process helps vaporise the remaining fuel and this speeds up the process. As the piston continues towards BDC, the exhaust valve will start to open around three-quarters of the way though the stroke. This allows the combustion gases to start to escape, forced out by the high pressure and temperature within the cylinder.

As already noted, the speed at which a diesel engine will run is determined mainly by the amount of fuel available. The high compression ratio used in the engines means that there will always be sufficient oxygen available within the cylinder to burn all the fuel, so this will never be a limiting factor. Primary speed control takes place by controlling the amount of fuel available to the cylinders. However speed is likely to vary for any fixed rate of fuel feed as engine conditions

[3]Timing is usually measured in terms of the rotational position of the crankshaft. Injection of fuel will start around 28° before the piston reaches TDC and ends at 3° after TDC.

change, depending upon the engine load, engine temperature and other factors. In order to maintain constant speed the engines are equipped with a feedback system that uses the engine speed to help modulate the fuel feed. In most diesel engines built before the 1980s this was carried out by a mechanical device called a governor. Modern engines use an electronic control unit which monitors and controls all engine parameters.

TWO-STROKE DIESEL ENGINES

Diesel engines are normally classified by their speed. Slow-speed engines operate at up to 275 rpm, medium-speed diesel engines have speeds of 275–1000 rpm and high-speed diesel engines operate at above 1000 rpm. These latter are the most common diesel engines and can be found in many applications including vehicles and smaller power generation applications. While the higher speed diesel engines are usually four stroke, many of the very large, slow-speed engines are two-stroke engines.

The two-stroke engine has a higher power to weight ratio than a four stroke because it has one power stroke for every revolution of the shaft rather than one for every two revolutions in the four-stroke engine. In small engines the two-stroke cycle is usually considered less efficient but the large slow-speed diesel engines are as efficient or more efficient than their four-stroke cousins.

Large two-stroke engines are extremely heavy and this results in their slow-speed engines often operate at 100 rpm or less. They have tall cylinders and this permits a long piston stroke. A long stroke combined with the slow-engine speed enables more precise control over combustion and this in turn enables a more efficient engine to be built. Very large two-stroke engines are also relatively more powerful than a four stroke; a large two-stroke engine is roughly 80% more powerful than a similarly sized four stroke. (This advantage reduces as the size of the engine decreases.)

Many large two-stroke diesel engines are designed for ship propulsion. In this application the slow speed means that the engine crankshaft can be connected directly to the propeller. In power applications the shaft is geared up to drive a generator.

One of the advantages of large two-stroke engines is their simplicity. The engines have no valves and so requires fewer parts. Another major advantage is the ability of these engines to burn heavy fuel oil with high efficiency. This heavy or residual fuel oil is the fraction left after all the more valuable parts of crude oil have been removed by fractional distillation. It is a cheap fuel but extremely dirty so plants that burn it need extensive emission control facilities. This can be cost-effective in a large plant whereas it might not be in a small diesel engine generating unit.

The size of these engines – the largest can have power generating capacities of up to 80 MW – means that it is also cost-effective to add a bottoming cycle. This involves using waste heat from the engine to generate steam which is then exploited to drive a small steam turbine, creating a combined cycle power plant. This can add a few percentage points to the overall efficiency in a large installation. This is illustrated in Fig. 5.3 which compares the thermal efficiencies of medium- and low (slow)-speed diesel power plants. As the curves in the figure indicate, slow-speed diesel engines are the most efficient with potential efficiencies of up to 60% when operated in combined cycle mode. However practical engine efficiencies are lower than this with the best achieved around 52%.

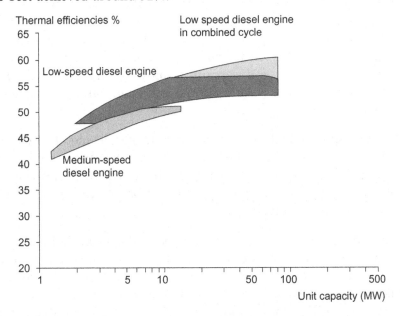

Figure 5.3 Comparative efficiency of low-speed diesel engines. Source: MAN Diesel and Turbo.

POWER GENERATION APPLICATIONS OF DIESEL ENGINES

Diesel engines are one of the most common reciprocating engines for use in power generation applications. High-speed diesel engines are often used as emergency and backup generators to provide power during grid outages. The ability of a diesel engine to start rapidly, often in less than 10 s, makes them particularly attractive in this respect. Where even faster cut-in is required, a diesel engine can be combined with a fast-acting energy storage system such as a supercapacitor or flywheel. Similar high-speed diesel engines are also used to provide power to remote communities that are not grid connected. The growth of renewable generation based on wind and solar power has reduced the reliance on the engines but modern off-grid systems often incorporate a diesel engine as backup for renewable generation to provide a reliable supply.

Medium-speed diesel engines can also be used to both backup supply and supply power to remote communities. However these engines are larger and more expensive than the high-speed engines so economic considerations become important. This type of engine is often used to provide power for industrial units that require their own power supply or cannot afford to lose grid power. Medium-speed engines can also be used for distributed generation applications where they provide power into a distribution system close to consumers. Medium-speed engines can be used for grid support too, particularly to provide power when intermittent, grid-connected renewable power is not available. However this type of grid support is most common in developed nations where there are strict emission codes and natural gas engines are often preferred.

Medium-speed diesel engines are easy to install and they can be used to provide base-load power in developing countries and to reinforce weak grid supplies. Capacity can be added by increasing the number of engines and the power plants can be relocated if necessary, making them very flexible.

Slow-speed diesel engines are the largest of the diesel fleet and these are usually used for base-load duty. The engines are particularly attractive when there is a source of heavy fuel oil because it provides an economical source of power. However the engines can also be used in a grid support role or in a situation where the power demand fluctuates.

One of the advantages of low-speed engines is their part load efficiency which varies very little over a range of outputs between 50% and 100% load. Whereas gas and steam turbines outputs tend to fall as the load decreases, that of a low-speed diesel engine is relatively flat over a similar load range. Efficiency can fall at lower loads but where this is an issue, plants can be built with several engines, with one or more taken out of service as load falls in order to raise the load and efficiency of the remaining engine(s).

CHAPTER 6

Stirling Engines and Free Piston Engines

In addition to conventional spark ignition and compression ignition engines there are a range of other engines types that rely on reciprocating pistons to generate power or work. Two important groups are Stirling engines and free piston engines.[1]

The Stirling engine is an external combustion engine in which the piston or pistons and the working fluid are sealed and isolated from the atmosphere. Heat and cooling are applied to the exterior of the engine to drive the engine cycle. This has the advantage that the engine can operate with a variety of different fuels and heat sources, although in practice engines are usually designed for specific fuels and applications. The Stirling engine concept is simple but realising it in a practical engine is complex. There are a wide range of Stirling engine designs but most can be understood in terms of one or two generic types. Most are crankshaft engines that use pistons mechanically linked to a crankshaft to provide power and to control the motion of the pistons. However some are a type of free piston engine. These do not use a crankshaft to control the movement of the piston(s).

Free piston engines are reciprocating engines in which the piston is not attached to any form of mechanical linkage such as a crankshaft in order to extract power. Instead the movement of the piston is normally controlled by the air pressure on opposing sides of the piston. As with Stirling engines, there are a variety of configurations for these engines. Since there is no crankshaft, power cannot be extracted from a free piston engine through a rotating shaft so alternatives are required. Some use linear generators attached to the pistons to produce electricity. Others act as compressors, with the high-pressure exhaust gases used to drive a turbine.

Stirling engines have a long history and the engines have been developed for a variety of power applications, from solar power to

---

[1]These two groupings overlap because some Stirling engines are free piston engines.

Piston Engine-Based Power Plants. DOI: https://doi.org/10.1016/B978-0-12-812904-3.00006-9

domestic heat and power. Free piston engines have had limited application since their invention but they are attracting more interest in the 21st century as potential power sources in hybrid vehicles.

THE STIRLING ENGINE

The Stirling engine is a heat engine that converts heat into mechanical energy. It does this by exploiting a heat source and a heat sink to alternately cause a gas, the working fluid, to expand and contract. The actual physical process that the engine exploits is the transfer of heat from the heat source to the heat sink and the working fluid, the gas, is the medium through which this transfer occurs. It is probably the closest approach in a practical engine to the idealised Carnot heat engine proposed by Sadi Carnot in the early 19th century.

The engine was designed by a Scottish Presbyterian minister, Robert Stirling, who received his first patent in 1816. The original Stirling engines used air within the cylinders and were called air engines but modern Stirling engines usually employ helium or hydrogen since these gases can absorb and release heat rapidly. The operation of early Stirling engines, which require high temperatures to operate efficiently, was limited by the materials available during the 19th century.

There are many configurations for Stirling engines, including Alpha, Beta, Gamma and free piston variants. All of these configurations require two pistons to function – although this is not always immediately apparent from their designs. The engine is a sealed unit so that the working fluid and pistons are isolated from the atmosphere. Movement of linked pistons cycle this working fluid from the hot part of the engine, where an external heat source provides heat to the fluid at constant temperature, and an external cold sink where heat is extracted from the engine. The engine cycle is reversible and if a Stirling engine is driven mechanically it can take heat energy from the cold end and move it to the hot end, essentially acting as a refrigerator.

STIRLING ENGINE CYCLE

The Stirling engine working fluid passes through four stages during its cycle: cooling, compression, heating and expansion. These four are similar to the same four stages that characterise all heat engine cycles and can be compared to the intake, compression, power and exhaust phases of an internal combustion engine. During these stages the gaseous working fluid behaves according to the gas laws that relate volume, pressure and temperature.

Fig. 6.1 shows a simplified Alpha Stirling engine at different stages of its cycle. As the figures indicate, the engine has two linked pistons which drive the working fluid, the gas, between a hot source and a cold sink. The working fluid in most modern Stirling engines is either hydrogen or helium which are good absorbers of heat. However both gases diffuse easily through seals and can be difficult to contain. Some recent machines have used nitrogen instead although its thermal properties are not as good as those of the other two gases. The gas inside the engine is highly compressed, often to 20 MPa, or 200 times atmospheric pressure.

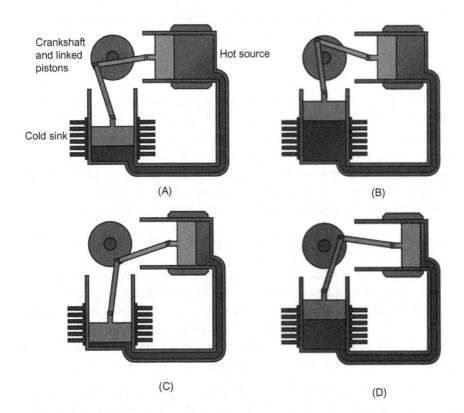

Figure 6.1 The Alpha Stirling engine cycle. Source: Wikimedia.

An Alpha Stirling engine has two synchronised pistons that oscillate, but with a time lag so that they do not move in or out either exactly together or exactly in opposition. This is important for the operation of the engine. Fig. 6.1A represents an Alpha engine at the point in its cycle when most of the working fluid is inside the hot cylinder and is being heated by the hot energy source. At the same time the piston in the cold cylinder is close to its minimum position and little working fluid is in contact with the cold sink. As heat passes into the gas from the hot source, the working fluid pressure increases and it expands, placing an additional force on the cold piston which starts to move so that the volume in that cylinder increases. Meanwhile the lag between the pistons means that the hot cylinder piston does not move significantly at this point in the cycle.

As the cycle continues the cold cylinder piston moves out, and the volume of the cold cylinder expands until the combined volume in the hot and cold cylinders reaches it maximum (Fig. 6.1B). At this stage the maximum amount of heat is being extracted from the cold cylinder and the mechanical linkage ensures that the hot piston now starts to move in, reducing the volume of the hot cylinder and forcing more of the gas into the cold region.

Now the amount of gas in the hot region approaches the minimum (Fig. 6.1C) and the extraction of heat from the cold cylinder starts to cause the gas volume to contract so that the cold cylinder piston starts to move down, reducing the volume of the cold cylinder too. As this continues the system reaches its minimum volume, as shown in Fig. 6.1D. At this stage the mechanical linkage sets the hot piston in motion, expanding the hot cylinder volume and the cycle begins once more.

One key element of the Stirling engine is missing from these simplified figures, a regenerator. This is a heat absorbing device that sits inside the tube connecting the hot and cold cylinders of the engine. The regenerator captures some of the heat from the gas that would otherwise pass out through the cold sink as the gas flows in that direction and then returns it to the gas as it passes back into the hot cylinder. This helps improve the efficiency of the engine. A schematic of an Alpha Stirling engine with a regenerator is shown in Fig. 6.2.

The other main category of Stirling engine is the Beta Stirling engine. This has two pistons contained within a single cylinder, with a

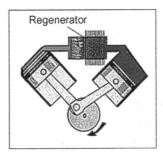

Figure 6.2 Alpha Stirling engine with regenerator. Source: Wikimedia.

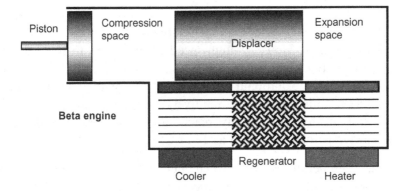

Figure 6.3 Beta Stirling engine. Source: Wikimedia.

hot source at one end and a cold sink at the other. One piston is a conventional power piston and seals the chamber to the atmosphere. The second is a loose fit within the cylinder so that as it oscillates the working fluid can move past it from bottom to the top of the cylinder and vice versa. A simplified version is shown in Fig. 6.3. The stages of the cycle are essentially identical to those of an Alpha engine.

In addition to these two main types there are a wide range of other Stirling engine designs. These include a gamma Stirling engine – a variation of the Beta engine but with a separate cylinder to the second piston, a flat Stirling engine and a number of free piston Stirling engine designs.

ENGINE TECHNOLOGY

The Stirling engines differ in several ways from a conventional internal combustion engine and have different mechanical and material requirements. The engines operate at high pressure and this imposes severe constraints, particularly on materials for engine construction and for piston seals. In order to resist the forces inside the engines due to this high pressure, Stirling engines are relatively heavy compared to similar internal combustion engines. In addition, the use of gases such as hydrogen and helium makes sealing the gas inside the engine, while allowing piston movement and lubrication, technically challenging. Both of these gases have small molecules that can diffuse easily so sealing the engines requires specialist materials and designs. Some engine designers have experimented with nitrogen or air as replacements, but these do not have the same advantageous thermal properties of either hydrogen or helium.

Many Stirling engines operate at relatively high speed, providing a high power density. However high-speed engines tend to have shorter lives, with greater wear, than slower speed engines. Larger engines tend to have lower rotational speeds by virtue of their greater mass. Stirling engines also operate at relatively high temperatures, with the heat source in engines typically reaching around 700°C while some operate at temperatures as high as 1000°C.

Higher temperatures lead to higher thermodynamic efficiency. Some commercially available engines have achieved up to 30% efficiency and in principle 50% efficiency is feasible. However the high temperature gradient in an engine can lead to thermal stress between the hot and cold sections of the engine and high-performance materials are required to manage this.

Another technological challenge is to achieve fast heat transfer from the heat source to the working fluid in the hot cylinder and from the cold cylinder into the heat sink. The choice of hydrogen or helium as the working fluid helps with heat transfer but the heat transfer elements of the engine body itself must also be extremely efficient at conducting heat. The design of an effective regenerator can also be challenging. As with the heat transfer elements, this must be able to absorb and release heat rapidly and effectively. Most regenerators are made from some form of fine mesh or wire construction that presents a high surface area for heat transfer as the working fluid passes through it. However this can get blocked if there are any particles within the working fluid.

Another issue with the Stirling engine is that it will not start immediately but requires a time to warm up. While this applies to many engines it tends to be more of a problem with external combustion engines. Steam engines, which are also external combustion engines, require extended start-up periods too.

ENERGY SOURCES

One of the attractions of the Stirling engine is that the heat energy is applied externally. Thus the energy can, in theory, be derived from any heat source. Unlike an internal combustion engine, heat is supplied continuously to the heat absorber of the engine and this makes the engine design simpler than that of a conventional internal combustion engine which requires the repetitive release of energy.

The heat energy can be supplied from a combustion fuel. This can be one of the many fossil fuels available; coal, oil and gas can all be used, although these are unlikely to be the main energy source for a Stirling engine power system. However other combustion fuels including biogas and some types of biomass can be used.

A Stirling engine can also operate on waste heat. This could be exhaust heat from a power plant or it could be heat released during an industrial process. The engines are more efficient, the higher the temperature of the heat supply, so a high-quality heat source will normally be needed to make the engine economical.

Another potential source of heat is nuclear power. While a Stirling engine is unlikely to be used in a terrestrial nuclear power plant today, they could potentially offer a means of generating power from a nuclear source in space. For the future there is some interest in Stirling engines in terrestrial nuclear power plants because they would eliminate the need for a heat transfer system and could greatly simplify the design. This would be particularly true of advanced nuclear power plants that use exotic heat transfer fluids such as molten sodium. Geothermal power is also a potential heat source. However this tends to be a low-temperature source which can limit efficiency.

Another important heat source is solar energy. This can be focussed using optical systems to provide a very high-temperature energy source and this has been successfully used to drive Stirling engines.

THE APPLICATION OF STIRLING ENGINES FOR POWER GENERATION

Stirling engines have been used to exploit solar energy and for biomass applications. However their use is not widespread. Typical engines sizes under development and in use range from 1 to 150 kW. In solar thermal applications a Stirling engine could theoretically achieve close to 40% energy conversion efficiency. The best so far recorded is 32% which is still high for solar conversion.

The most common solar application is in a solar dish power system. This type of power plant has a large reflecting dish, up to 25 m in diameter which focusses the sun's energy onto an absorber placed at the focal point. The heat energy collected at this point is transferred to the hot side of a Stirling engine to provide the thermal input to drive the engine. The cold side is provided by air at ambient temperature. Solar systems based on dishes are relatively small with generating capacities of 25–50 kW. A 10 m dish can provide energy for a 25 kW Stirling engine.

Another area for which Stirling engines have recently been developed is small combined heat and power. These systems are usually aimed at domestic electricity and heat supply with the heat energy provided from natural gas, although other sources are possible. In these systems the Stirling engine provides electricity while the heat that is not used by the engine is used to heat hot water for domestic hot water and space heating systems.

FREE PISTON ENGINES

Free piston engines are so called because the pistons in the engines move linearly inside their cylinders without any mechanical constraint such as a crankshaft. This simplifies engine design and the engines are potentially more efficient because they do not have frictional and other losses associated with more normal mechanical linkage designed to convert the reciprocating motion into rotary motion. However it is more difficult to extract energy from a free piston engine.

The first modern free piston engine was designed by the Argentine engineer Raúl Pateras Pescara. The first design was for a compressor developed and marketed by Pescara Auto-compressor Company which was launched in 1933. He later produced a generator based on the free piston principle. The development of this type of engine continued with their use as gas generators in which the high-pressure exhaust from the engine is used to drive a gas turbine. Free piston engines have also been coupled with some form of linear generator that can exploit the back-and-fro movement of the piston, rather than the more normal rotary motion, to provide an electrical output.

TYPES OF FREE PISTON ENGINE

The simplest type of free piston engine has a single cylinder that operates in a manner identical to the cylinder of a two-stroke spark ignition engine. Air and fuel are mixed and fed into the cylinder, the mixture is compressed and then ignited, providing a power stroke and at the same time exhausting the combustion gases and the cycle is repeated. In the free piston version, the piston is connected not to a crankshaft but to a bounce chamber, a second piston that is sealed inside a cylinder containing air or another gas. This is shown schematically in Fig. 6.4. During the power stroke the main engine piston forces the bounce chamber piston to compress the gas inside the chamber. At the end of the power stroke the gas inside the bounce chamber is under high pressure and can force the engine piston to return to the top of its chamber, starting the compression and ignition stroke again.

This type of free piston engine is extremely simple. The pressurisation of the gas inside the bounce chamber can be exploited to provide a form of hydraulic drive. Alternatively it will act simply to return the piston while a linear generator exploits the back-and-fro motion of the piston to generate electricity.

Figure 6.4 Single-piston, free piston engine. Source: Newcastle University, UK[2].

[2]A review of free-piston engine history and applications. R. Mikalsen, A.P. Roskilly Sir Joseph Swan Institute for Energy Research, Newcastle University, Newcastle upon Tyne, NE1 7RU, United Kingdom.

Figure 6.5 Dual-piston, free piston engine. Source: Newcastle University, UK[3]

A second common type of free piston engine is the dual-piston engine. This has two internal combustion engine cylinders each with its own piston. However these pistons are connected, back to back so that as one piston moves through its power stroke it compresses the contents of the second piston (as if it were a bounce chamber), and vice versa (Fig. 6.5). This type of design eliminates the need for an actual bouncer chamber. However it requires extremely accurate control of the combustion cycles in each cylinder since each drives the other. Both the stroke length and the compression ratio in each of the cylinders is partly under the control of the second and this can lead to imbalances if synchronisation is not perfect. The engines are highly efficient, in principle. As with the single-piston version, the engine can be used either for hydraulic drive or to generate electricity via a linear generator.

A third type is the opposed-piston, free piston engine. This also has two pistons but they are inserted at either end of a single combustion chamber, or cylinder. The outside ends of each piston are attached through a rod to a piston inside a bounce chamber. In this case, ignition and firing of an air–fuel mixture in the cylinder causes the two pistons to move outwards, symmetrically. In order to ensure that synchronisation is perfectly maintained, there is usually a mechanical linkage controlling their relative motions. This is shown in Fig. 6.6 which presents a schematic of an opposed-piston, free piston engine. When the pistons reach the end of their strokes they are returned by the bounce chambers and the cycle is repeated. As a consequence of the two opposed pistons, this type of engine is vibration free. However it is more complex than other types of free piston engine.

An alternative approach to the use of free piston engines for power generation is the free piston gas generator. This exploits the performance of an opposed-piston, free piston engine to produce a flow of compressed gas which is supplied to the combustion chamber of a gas

---

[3]A review of free-piston engine history and applications. R. Mikalsen, A.P. Roskilly Sir Joseph Swan Institute for Energy Research, Newcastle University, Newcastle upon Tyne, NE1 7RU, United Kingdom.

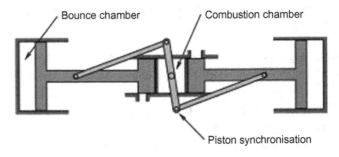

Figure 6.6 Opposed-piston, free piston engine. Source: Newcastle University, UK[4]

turbine. In this application the free piston gas generator replaces the compressor of the gas turbine. The free piston device can provide a high compression ratio and the devices were of considerable interest in the middle of the 20th century but as gas turbines developed through the latter part of the century, interest waned.

LINEAR GENERATORS

In order to generate electricity from a free piston engine, some form of linear generator is required. There has been a considerable amount of interest in free piston linear generators during the 21st century because they could be used as range extenders for electric vehicles. Most of these devices integrate the free piston engine and the generator in a single device. For vehicle applications, the idea is to use the engine to provide an efficient source of electricity from combustion of a standby supply of combustion fuel in case the vehicle battery runs low.

A number of configurations have been explored. For example, one design developed at the Sandia National Laboratory in the United States utilises magnetised pistons, with coils around the cylinder to produce an electrical output.[5] Most use opposed-piston engines with a variety of ignition options including spark ignition, compression ignition and homogeneous charge compression ignition in which an air–fuel mixture is compressed to the point of spontaneous ignition, as in a diesel engine.

[4]A review of free-piston engine history and applications. R. Mikalsen, A.P. Roskilly Sir Joseph Swan Institute for Energy Research, Newcastle University, Newcastle upon Tyne, NE1 7RU, United Kingdom.
[5]http://crf.sandia.gov/free-piston-engines-a-possible-route-to-hybrid-electric-vehicles/

APPLICATIONS OF FREE PISTON ENGINES TO POWER GENERATION

Free piston engines can be used to generate electric power either by coupling with a linear generator or by using the engine as a gas generator for a gas turbine. However while many configurations have been explored, there are no major commercial applications. The main interest today is from the automotive industry where they have a potential use as range extenders. If these are developed successfully the engines might become both cheap and reliable enough for other uses such as small distributed power generation or combined heat and power applications.

CHAPTER 7

Piston Engine Cogeneration and Combined Cycles

Reciprocating engines can be relatively efficient — at least compared to other types of heat engine — at converting fuel into electricity. The best diesel engines can achieve 50% efficiency. Similar sized gas turbines are several percentage points less efficient though the largest can match this. However even if an engine is 50% efficient it means that roughly 50% or more of the energy in the fuel is not converted into electrical power and instead emerges as heat.

If there is a use for it, this heat can be captured either in the form of hot water or as steam. This can raise the overall efficiency of energy usage to 80% or more. There are two broad ways in which the waste energy can be exploited. The first is to capture and use the heat directly. This will often provide space heating or hot water but the heat from some parts of the engine can be of high enough quality to generate low- or medium-pressure steam which can be used in some industrial and commercial processes. Engines are necessarily equipped with cooling systems of various types to ensure engine components do not overheat and these can be exploited easily to provide a heat output.

The second approach is to use the waste heat as an additional source of electric power. This requires a second (bottoming[1]) turbine generator to be added, one that can use the waste heat to drive its cycle. Some large engines are equipped with steam generators and small steam turbines to create diesel engine combined cycle plants. This is normally only economical for the largest of installations. For smaller engines it might be cost-effective to use an organic Rankine cycle (ORC) turbine that can exploit lower grade heat economically. The use of ORC bottoming cycles is not widespread but interest is growing.

---

[1]A bottoming cycle is so called because it takes rejected energy after it has emerged from the 'top' cycle which in this case is a diesel engine.

Piston Engine-Based Power Plants. DOI: https://doi.org/10.1016/B978-0-12-812904-3.00007-0

RECIPROCATING ENGINE COGENERATION SYSTEMS

When an internal combustion engine is used to generate electricity, a large part of the energy supplied to the engine in the form of fuel emerges as heat in the exhaust from the engine or is dumped to the atmosphere by engine cooling systems. If this heat can be captured it can be utilised for space heating or for heating water, potentially making the energy usage much more efficient.

The efficiency of piston engine-based power generation varies from 25% for small engines to close to 50% for the very largest engines. This means that between 50% and 75% of the fuel energy actually emerges as waste heat. If this heat can be captured it can be utilised in a cogeneration system. This type of system is often also called a combined heat and power (CHP) system and it can raise the efficiency of fuel usage to 80% or more. The overall efficiency will depend on engine type and size.

Heat energy capture can be applied to all the main types of reciprocating engines including spark ignition engines and compression ignition engines. The environmental demands in the 21st century mean that in many cases the engines that are used will be gas engines using spark ignition and natural gas as fuel but all these internal combustion engines share similar components, whatever the fuel or ignition type, and all can be adapted for cogeneration.

In the past reciprocating engine CHP systems have only been cost-effective in installations where the electrical generating capacity of the engine is 50 kW or above. Some smaller systems have been produced but at lower electrical output the electrical efficiency falls and maintenance costs rise. Small reciprocating engines are normally mass-produced and they are not designed for continuous base-load operation. In the last 5 years there has been considerable development directed at improving the efficiency and longevity of these smaller engines and this has led to much smaller systems, in the range of 1−5 kW, being marketed for domestic and small commercial applications. This is a small but expanding market.

The larger reciprocating engine CHP systems will normally be used to supply base-load electricity, often with the ability to sell surplus power to a local grid. These engine systems are usually supplied as a complete package. In the case of a gas engine, all that is required is to connect hot water and a gas supply and the system is ready to go. However large systems, particularly where steam is being generated for process use, may be customised to each installation.

The electrical generation efficiency of a reciprocating engine usually increases with engine size. Typically, an engine with 100 kW electrical output will be expected to have an electrical generating efficiency of around 27% while for a 1 MW system efficiency rises to 37% and at 10 MW efficiency approaches 42%.[2] The smaller engines also typically have a higher exhaust gas temperature than the larger engines. The lower electrical efficiency is not necessarily a handicap in a CHP system. It simply means that there is relatively more heat energy available to capture and reuse: For a small engine the heat may account for over 70% of the total energy input while for a larger engine it will be less than 60%. So, for a 100 kW engine, the heat recovery might be close to 200 kW — a heat to electricity ratio of 2:1 — while for a 10 MW engine the total heat recovery may only be 8 MW — a heat to electricity ratio of 0.8:1.

While these figures cover off-the-shelf systems, it is possible to adapt engine design to meet different thermal and electrical requirements. So, for example, a larger engine might be modified so that it was less efficient, electrically, but produced more heat that could be captured and used in some industrial or commercial process.

ENGINE HEAT SOURCES

There are four primary sources of waste heat in an internal combustion engine: the engine exhaust, the engine case water cooling system, the lubrication (lube) oil water cooling system and, where one is fitted, the turbocharger cooling[3] system. Each of these can be used as a source of heat in a reciprocating engine cogeneration system. Fig. 7.1 shows the first three of these, schematically, for a natural gas engine cogeneration system.

---

[2]Figures are taken from the Energy Solutions Center website: http://understandingchp.com/chp-applications-guide/4-chp-technologies/.
[3]A turbocharger is sometimes used to compress air before it is admitted into the cylinder of an internal combustion engine. This can lead to improved performance by generating greater power from the engine.

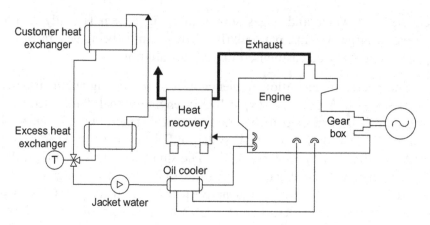

Figure 7.1 Heat sources for a reciprocating gas engine CHP system.

The exhaust gas contains up to one-third of the fuel energy and 30%–50% of the total waste heat from the engine. Exhaust heat is not normally captured in conventional engines but it is straightforward to fit a heat recovery system to the exhaust of an engine if the heat is required. The exhaust temperature is typically between 370°C and 540°C. This is sufficiently high that it can be used to generate medium-pressure heat if required, with a maximum pressure of around 28 bar. Otherwise it can be used to generate hot water. Engine exhaust gases have also been used directly for drying in some applications.

The main engine case cooling system can capture up to 30% of the total energy input. Cooling water exits the cooling system at up to 95°C.[4] In a cogeneration system this will be passed through a heat exchanger to provide a source of hot water. Engine oil and turbocharger cooling systems will provide additional energy that can also be used to supply hot water. The engine cooling jacket and oil cooling system will typically provide 45%–55% of the total waste heat recovery from an engine system.

If all the heat from the exhaust and the cooling systems of an engine is exploited, around 70%–80% of the fuel energy can be used. However this can generally only be fully exploited when there is a need for hot water. Overall efficiency will also depend on the duty

---

[4]It may be hotter if the cooling system is pressurised.

cycle of the engine. Most reciprocating engines will show little fall in overall efficiency when the electrical load on the engine falls from 100% to 50% but if it falls lower than 50% then efficiency will start to fall more sharply. Engine and waste heat temperatures are likely to fall too, and so a wide daily variation in load is likely to have an impact on the effectiveness of the CHP system.

Since cooling systems are fitted to internal combustion engines whether the waste heat is exploited or not, the use of these systems in CHP applications offers a logical extension of their application. Cogeneration systems based on small engines can provide power, space heating and hot water for commercial offices while large engines can produce power and low-grade process heat for small industrial operations. The economics of these systems can be quite favourable where there is a use for the waste heat. As a consequence the cogeneration market, particularly for small systems, is buoyant and is likely to become more so if fuel costs continue to rise.

STIRLING ENGINE CHP SYSTEMS

From a CHP perspective, Stirling engines offer a different challenge to the conventional reciprocating engine because these engines do not normally have extensive cooling systems. However there is a flow of waste heat from the engine and this can be captured and used. Most commercial Stirling engines are relatively small, with typical sizes from 1 to 25 kW, although larger engines, up to 100 kW, are likely to become available in the future. These small engines have a low power to heat ratio so there is significant amount of heat available and this, together with their size, makes them particularly suitable for domestic or small commercial CHP systems. Residential CHP systems have been built around the Stirling engine and this could provide an important future market if technology costs can be reduced sufficiently. Stirling engines are not very good at load following, so domestic systems are likely to be sized so that the engine can run continuously. Depending upon design considerations, the engine might be small (typically around 1 kW) and only supply a part of the domestic supply, or it might be relatively large (3–5 KW) and supply most of the power needed by the household, with surplus being exported to the grid.

COMBINED CYCLE

The waste heat from the exhaust of an internal combustion engine is generally hot enough to generate medium-pressure steam at up to 400°C and 15–30 bar (more often at the low end of this range). In the case of small engine installations, steam production is not normally an economical option unless there is a local use for low-quality steam. In the case of a very large diesel installation, however, the engine exhaust can be used to generate steam in a boiler, steam which can drive a steam turbine to produce additional energy. This forms the core of a diesel engine-based combined cycle plant.

Diesel engine CHP systems are rare because they are generally only economical on very large engines. Typical of this sort of application is a generating plant which was installed in Macau in 1987. This plant was equipped with a slow-speed diesel engine with a capacity of 24.4 MW. The engine exhaust was fitted with a waste heat boiler and steam turbine which could generate an additional 1.34 MW when the engine was operating at full power, thus contributing around 5% of the plant output. As a result of this and other measures a fuel-to-electricity conversion efficiency of close to 50% was achieved. More modern installations aim to exceed 50% electrical efficiency.

Large engines of this type are frequently derived from marine engines and the original engines upon which they are based are not normally optimised for combined cycle operation. In particular, the cooling system is designed to keep the engine as cool as possible. For best combined cycle performance, however, it is preferable to run the engine as hot as possible because the higher the exhaust gas temperature, the more efficient the steam turbine cycle. High-temperature operation can also improve engine efficiency because the potential thermodynamic efficiency will increase with operating temperature.

Combined cycle performance of a large diesel engine can therefore be improved by modifying engine components such that they can operate continuously at a higher temperature. Such modifications may require more expensive materials capable of withstanding the more extreme conditions. For example, the top of the piston may be made from an alloy that allows it to remain uncooled while exhaust valves are treated with advanced coatings able to resist the high exhaust gas temperature.

These modifications allow a higher temperature exhaust which can be used to generate higher quality steam to drive a steam turbine.

With these measures it may be possible to achieve a fuel to electricity conversion efficiency of close to 55%. This was the target for efficiency for a plant in Wasa, Finland, installed in 1998. The plant has two 17 MW diesel engines and a single steam turbine. Efficiency in this case was improved by using seawater cooling for the steam turbine condenser. The additional expense of the waste heat recovery and steam turbine will generally only prove cost-effective if the engine is to be used for base-load operation.

ORGANIC RANKINE CYCLE

An organic Rankine[5] cycle turbine is a small turbine that is identical in design to a steam turbine but which uses a low boiling point organic fluid as its working fluid instead of water and steam. The turbine is packaged into a closed cycle system with a condenser and heat exchanger. Waste heat is supplied to the heat exchanger and this vaporises the organic fluid. The vapour drives the small turbine with a condenser on its outlet to ensure a high pressure drop across the turbine. The re-condensed fluid is returned to the heat exchanger. The turbine, meanwhile, is connected to a generator which provides an electrical output.

The ORC has been used to exploit low-quality heat from a number of different sources. They have been used with low-temperature geothermal reservoirs and can be added as bottoming cycles for a range of different heat engine generating systems. The ORC would not normally be considered a cost-effective addition to a reciprocating engine system. Recently, however, small mass produced and packaged ORC systems that can exploit any waste heat source have become available. These can relatively easily be adapted for use with a reciprocating engine. One company that has developed an ORC system adapted for a 1 MW diesel generator package has claimed an improvement in fuel efficiency of 10%–12% and a payback time of 2–3 years for a diesel-based plant.[6]

---

[5]The Rankine cycle describes the operation of the steam turbine as a heat engine.
[6]Efficiency Gains by Bottoming Reciprocating Engines with an ORC, Rob Emrich, ElectraTherm, 2015.

CHAPTER 8

The Environmental Impact of Reciprocating Engine Power Plants

Reciprocating engines, as a class, are one of the major sources of atmospheric pollution across the globe. Most engines burn fossil fuels such as gasoline or diesel and they generate a range of combustion products that are emitted along with the exhaust gases from the engine. They are also responsible for the release of large quantities of carbon dioxide into the atmosphere.

The largest part of global reciprocating engine emissions are from engines used for automotive applications. Most small automotive engines adopt limited emission control strategies. Cost prevents the application of more effective measures. While these limited controls will reduce emissions of the most damaging pollutants, they do nothing to control carbon dioxide emissions, which is why there is a push to develop vehicles that do not rely on internal combustion engines.

When reciprocating engines are used for stationary power, they generate a range of pollutants that are similar to those of automotive engines. These include nitrogen oxides, carbon monoxide, unburnt hydrocarbons and volatile organic compounds (VOCs), small particles (called particulates) and in the case of the largest diesel engines, sulphur dioxide. There are cost-effective strategies that can be used to control these emissions from stationary power plants and local regulations will normally require that they be applied. However there are none to tackle the carbon dioxide emissions which these engines produce.

Stationary engines will have other environmental effects too. There will be some heat emissions — these may be very small if heat energy is captured and reused — and reciprocating engines are noisy. Engines that are being used in urban environments will need extensive sound-proofing to combat the noise emissions. Very large engine-based power plants will require a source of cooling water and depending on the

Piston Engine-Based Power Plants. DOI: https://doi.org/10.1016/B978-0-12-812904-3.00008-2

type of engine there may be regular fuel deliveries which will add to local traffic movement. All these factors have to be taken into account when considering a new engine installation.

THE ORIGIN OF RECIPROCATING ENGINE EMISSIONS

Piston engines in stationary power units virtually all utilise a fuel that is burnt to release energy which is then exploited in the heat engine to provide electrical power. The energy source is most commonly a fossil fuel such as gasoline, diesel or natural gas, although some engines burn sustainable organic fuels such as bio-diesel and ethanol. The combustion of all these fuels, including the biofuels, produces a range of chemical by-products, many of which are toxic or harmful. If these are not removed or rendered harmless they will be released into the environment along with the exhaust gases from the combustion process. In the case of internal combustion engines the main emissions are nitrogen oxides, carbon monoxide, VOCs and particulates. Large diesel engines burning heavy diesel fuel may also produce some sulphur dioxide. The emissions of all these can be reduced by application of appropriate technology.

In addition, the combustion process will produce carbon dioxide. It is unlikely there will ever be a cost-effective method of removing the carbon dioxide from the exhaust of a reciprocating engine. The only solution is to burn a carbon neutral biofuel such as bio-diesel or ethanol.

Nitrogen oxides, NO_x, are formed during combustion, primarily by a reaction between nitrogen and oxygen in the air mixed with the fuel. NO_x can also be produced from nitrogen contained in fuel but there is little nitrogen in the liquid and gaseous fuels burnt by most reciprocating engines.

The main NO_x product in an internal combustion engine is nitric oxide, NO. The reactions that lead to NO take place more rapidly at higher temperatures when oxygen and nitrogen in air are at their most reactive. In lean-burn gas engines, spark ignition engines where the fuel is burned with an excess of air, temperatures can be kept low enough to that nitrogen oxide emissions are within local limits. The engine timing can also be adjusted to reduce production of NO_x but this may effect engine efficiency. The diesel cycle depends on relatively high temperatures to operate and as a consequence of this produces relatively high levels of nitrogen oxides. Table 8.1 compares emissions from the two types of engine.

When the fuel in an internal combustion engine is not completely burned the exhaust will contain both carbon monoxide and some unburnt hydrocarbons. This can happen in any engine, particularly when the load is changing and during start-up and shutdown. Carbon monoxide is hazardous at low levels and its emissions are regulated like those of NO_x. Unburnt hydrocarbons are classified as VOCs and their emissions are also controlled by legislation. A rich fuel–air mixture will produce more of both (as well as more NO_x) than a lean mixture because with less oxygen available, there will be more fuel that remains incompletely combusted. On the other hand, the combustion temperature in an engine burning a rich fuel–air mixture is higher than with a lean mixture. This leads to greater efficiency because a heat engine can extract more power from the engine if it has a higher peak combustion temperature. In order to reduce emissions, efficiency must therefore be compromised.

Particulates are another class of emissions. These small particles result from incomplete combustion too; they are essentially a type of soot. The smaller the particles, the more dangerous they become, with the smallest able to enter the blood stream through the lungs if breathed in with air. Diesel engines produce significantly higher levels of particulates than natural gas engines or gasoline engines. Liquid fuels, particularly heavy fuel, may also produce particles derived from ash and metallic additives.

Natural gas contains negligible quantities of sulphur so gas engines produce no sulphur dioxide. Gasoline contains no sulphur either. Diesel fuels can contain sulphur. Small- and medium-sized diesel engines generally burn lighter diesel fuels which contain little sulphur. Larger engines can burn heavy residual oils which are comparatively

Table 8.1 Emissions of Nitrogen Oxides From Internal Combustion Engines		
	Emission (ppmV)	Emissions (g/kWh)
High-speed and medium-speed diesel engines	450–1800	7–20
Spark ignition natural gas engine	45–150	1–3
Source: US Environmental Protection Agency.		

Table 8.2 Range of Emissions From Diesel Engines	
Emission	Emission Range
Nitrogen oxides	50–2500 ppmV
Carbon monoxide	5–1500 ppmV
Particulate matter	0.1–0.25 g/m^3
VOC	20–400 ppmV
Sulphur dioxide	10–150 ppmV
Source: Nett Technologies.	

cheap but which often contain significant levels of sulphur. Since sulphur can damage the engine, it is normal to treat this type of fuel first to remove most of the sulphur. The range of emissions from diesel engines are shown in Table 8.2.

NITROGEN OXIDE EMISSIONS

The most significant exhaust emissions from a piston engine are nitrogen oxides. These can both cause and aggravate respiratory diseases such as asthma. Continual exposure to high levels can lead to serious respiratory problems. In addition, NO_x can react with other airborne pollutants to generate both particulates and ozone, both of which are themselves harmful to health. These reactions are also responsible for the haze above many cities. In the wider environment NO_x can react in the atmosphere to generate acid rain that can damage plants and trees and acidify lakes and rivers, killing wildlife that relies on them.

There are a number of strategies for reducing the emissions of NO_x. Engine modifications that reduce the combustion temperature of the fuel, such as the use of a lean fuel mixture and timing adjustments can provide the first step in reducing these emissions. Natural gas engines designed to burn a very lean fuel (excess air) provide the best performance, 45–150 ppmV or 1–3 g/kWh. Diesel engines present a greater problem because of the higher combustion temperature.

An additional technique that is being applied to internal combustion engines to reduce NO_x is exhaust gas recirculation. This involves taking some of the exhaust gas and mixing it with the air used to feed the engine. The effect is to reduce the overall oxygen concentration and thereby reduce the combustion temperature. This reduces NO_x production but will also reduce engine efficiency.

Such techniques improve emission performance and may be adequate for automotive applications. However they are unlikely to be sufficient for stationary engines for power generation. Depending on the regulatory regime and the engine some additional form of post-combustion NO_x emission control will usually be required.

For small gasoline engines a simple catalytic converter of the type used in automobiles is often the most effective solution. However this type of system cannot be used with diesel or with lean-burn engines. New catalysts for use with lean-burn engines are currently under development. Where a catalytic converter can be used, nitrogen oxide reduction is around 90% or more.

Automobile style catalytic converters are a relatively expensive means of reducing NO_x emissions. For large engines, the more economical alternative is to use a selective catalytic reduction (SCR) system and this system can be applied to both stationary engines and large transportation engines. SCR also employs a catalyst, but in conjunction with a chemical reagent, normally ammonia or urea, which is added to the exhaust gas stream before the emission control system. The reagent and the nitrogen oxides react on the catalyst and the nitrogen oxides are reduced back to nitrogen. This type of system will cut emissions by 80%–90%. However care has to be taken to balance the quantity of reagent added so that none emerges from the final exhaust to create a secondary emission problem.

CARBON MONOXIDE, VOCs AND PARTICULATES

The emission of carbon monoxide, VOCs and some particulate matter can be partially controlled by ensuring that the fuel is completely burnt within the engine. This is simplest in lean-burn engines but conditions within these engines does compromise efficiency. With all engines, careful control of engine conditions and electronic monitoring systems can help maintain engine conditions at their optimum level. Old engines as they become worn can burn lubrication oil, causing further particulate emissions.

For larger engines, and particularly for diesel engines, engine control systems will not maintain emissions sufficiently low to meet statutory emission standards. In this case an oxidation catalyst will be needed to treat the exhaust gases. When the hot gases pass over oxidation catalyst, carbon monoxide, unburnt hydrocarbons and carbon particles are oxidised by reacting with oxygen remaining in the exhaust gases, completing the combustion process and converting all the materials into carbon dioxide.

SULPHUR DIOXIDE

Sulphur emissions can be found in diesel engines that burn fuel containing sulphur. Many engines now burn low-sulphur fuels with less than 0.05% sulphur content. However some diesels and the heavy fuel oils that very large engines burn may contain significant amounts of sulphur. Heavy fuel, or residual oil, may contain as much as 3.5% sulphur. The best way of controlling sulphur emissions from internal combustion engines is to remove the sulphur from the fuel before use. However in the worst case a sulphur capture system can be fitted. This is likely to be similar to the scrubbing tower used in a coal-fired power plant but at much smaller scale. The use of such a system will add to both capital and maintenance costs and affects plant economics. It is only likely to be cost-effective in the very largest reciprocating engine-based power plants.

CARBON DIOXIDE

Internal combustion engines, in common with all heat engines that burn carbon-based fuel, generate carbon dioxide which is released into the atmosphere with the exhaust gases leaving the engine. The relative amount produced during electricity generation depends both on the fuel and on the efficiency of the engine. A large, high-efficiency diesel engine operating at close to 50% efficiency will produce significantly less carbon dioxide for each unit of electricity it generates than a small gasoline engine operating at perhaps 20% efficiency.

Currently the only way of effectively eliminating carbon dioxide emissions from such engines is to run them on a biofuel such as ethanol or bio-diesel that has been derived from plants. The principle here is that although the combustion of the fuel will still produce carbon dioxide, the re-growth of the plants that were used to produce the fuel will absorb the same amount of carbon dioxide from the atmosphere, so that for a full cycle of growth, fuel production and combustion, the net amount of carbon dioxide added to the atmosphere is zero.

Research is underway to develop systems to capture carbon dioxide from the exhaust of fossil fuel combustion plants and a variety of techniques are being explored based on some form of post-combustion capture. Whether such systems will ever be used extensively on reciprocating engines seems doubtful since the cost is likely to be prohibitive. For very large stationary power generating systems it might eventually be both technically feasible and economical. However the rate at which alternative renewable technologies are advancing may make such technology unnecessary.

ADDITIONAL ENVIRONMENT EFFECTS

In addition to the range of atmospheric emissions that are generated by the combustion of a fuel in a reciprocating engine, the installation of an engine system for power generation can lead to a range of other environmental ramifications. Most are low level but all need to be considered when such a project is under consideration.

One of the key factors to consider is noise. Reciprocating engines are extremely noisy mechanical devices and they produce an audible signature over a wide range of frequencies. This may not be a problem in an industrial environment but for a commercial or domestic installation it will represent a significant issue. Most commercial stationary power systems based on reciprocating engines are packaged with soundproofing to reduce the audible emissions to an acceptable level. However there will be cases where additional isolation measures are necessary.

Operating engines produce significant quantities of waste heat that will be absorbed in the local environment. In the case of cogeneration systems, the amount of heat waste may be as little as 20% of the total energy input from the fuel but for power-only systems it could be as much as 60%. This will lead to a local temperature rise that could affect the environment although the impact will generally be small.

Installation and decommissioning of a stationary power generation unit will cause some temporary disruption. The significance of this will depend on the size of the installation. Some types of engine may require regular fuel deliveries by road. There will also be regular maintenance interventions and there is always the danger of a spillage of fuel or lubricating oil which could enter local waterways.

CHAPTER 9

The Economics of Piston Engine Power Plants

The economics of power generation based on reciprocating engines depends to a large extent on the use to which the engine is to be put. Small electricity generators based on piston engines are relatively cheap: for intermittent use these may be ideal but they are not designed for continuous use and if exploited in this way will soon reveal their limitations. As a backup generator for a domestic household or as a portable source of electric power such a device may be ideal. For an engine of this type, the up-front or capital cost will be the most important factor determining choice.

Smaller engines that have been designed specifically for power generation tend to be much more expensive but equally much more reliable. These may be used for commercial backup systems in which case reliability and capital cost are likely to be the deciding factors. However many of these engines will also be used for distributed power or heat and power generation. Some recent small engines systems are aimed at domestic combined heat and power (CHP). The cost equation then becomes a matter of comparing the cost of electricity from an reciprocating generator to the cost of grid power, or to the cost of alternative distributed generation technologies such as micro turbines. The economics of a distributed generation CHP power system will be different to those of a unit that is for power generation alone. The cost of fuel will also begin to enter the economic equation.

Larger engines, of 100 kW or more, are often designed for base-load applications. Such engines are expected to be able to operate round the clock (but with regular maintenance intervention) for 20 years or more. These engines are likely to be used in commercial and municipal institutions as well as for industrial applications. Engines in this size range are also suitable for grid support. Cost is high but the relative cost will depend upon engine size. Larger engines tend to offer better economics and lower unit capital cost. Fuel efficiency is important too and the choice of fuel, and fuel cost, will be an important part of the equation.

Piston Engine-Based Power Plants. DOI: https://doi.org/10.1016/B978-0-12-812904-3.00009-4

Large engines-based plants may have generating capacities of several megawatts: For the largest slow-speed diesel plants the total capacity could be up to 80 MW. Plants of this size will require serious economic evaluation before they are constructed. This will often involve the use of an economic model called the levelized cost of electricity (LCOE) model, a lifetime cost model which allows a comparison to be made between different generating technologies. Using this model, the price of electricity from a range of potential power plant technologies can be compared, and a choice made based on cost.

COST OF ELECTRICITY

The cost of electricity from a power plant of any type depends on a range of factors. First there is the cost of building the power station and buying all the components needed for its construction. In addition, many power projects today are financed using loans so there will also be a cost associated with paying back the loan, with interest. Then there is the cost of operating and maintaining the plant over its lifetime, including fuel costs if the plant burns a fuel. Finally the overall cost equation should include the cost of decommissioning the power station once it is removed from service.

It would be possible to add up all these cost elements to provide a total cost of building and running the power station over its lifetime, including the cost of decommissioning, and then dividing this total by the total number of units of electricity that the power station actually produced over its lifetime. The result would be the real lifetime cost of electricity from the plant. Unfortunately such as calculation could only be completed once the power station was no longer in service. From a practical point of view, this would not be of much use. The point in time at which the cost-of-electricity calculation of this type is most needed is before the power station is built. This is when a decision is made to build a particular type of power plant, based normally on the technology that will offer the least cost electricity over its lifetime.

In order to get around this problem economists have devised a model that provides an estimate of the lifetime cost of electricity before the station is built. Of course, since the plant does not yet exist, the model requires that a large number of assumptions be made. In order to make this model as useful as possible, all future costs are also converted to the equivalent cost today by using a parameter known as the discount rate. The discount rate is almost the same as the interest rate and relates to the way in which the value of one unit of currency falls (most usually, but it could rise) in the future. This allows, for example, the cost of replacement of a plant component 20 years into the future to be converted into an equivalent cost today. The discount rate can also be applied to the cost of electricity from the power plant in 20 years' time.

The economic model is called the LCOE model. It contains a lot of assumptions and flaws but it is the most commonly used method available for estimating the cost of electricity from a new power plant. One particular problem is that the model does not take into account cost risks. For example the cost of natural gas can fluctuate widely so that it may be cheap to buy gas when a plant is built but 5 years later the cost is so high that operation of the plant is uneconomical. The level at which the discount rate is set can also be problematical. It is typical to use a discount rate of 5% and 10% in calculations. However in the middle of the second decade of the 21st century the actual interest rate is close to zero.

FUEL COSTS

One of the main cost elements for a piston engine power plant is the fuel needed to operate it. There will be cases, such as where the plant is intended to provide emergency backup, where the cost of fuel is a subsidiary consideration. In most cases, however, the fuel cost will help determine the optimum power plant configuration.

The main fuels for piston engine power plants, gasoline, natural gas and diesel, are all commodities whose costs are determined based on market demand. The cost of gasoline and diesel is generally determined by the global cost of oil. As this rises and falls, so does the cost of the fuel.

During the last 40 years cost of oil had fluctuated widely. For example the price of a barrel of Brent Crude in 1981 was $36. By 1986 it was down to $15/bbl, in 1998 it was $12/bbl but in 2011 and 2012 the average price was $111/barrel.[1] In 2016 the price was $44/bbl.

[1]BP Statistical Review of World Energy 2017.

When prices fluctuate by this much, the economics fluctuate too. It may be cost-effective to build a power plant using diesel fuel but by the time the plant enters service, the cost may be too high for it to be economical to operate. This level of risk must always be considered.

Natural gas has been subjected to similar fluctuations. The global cost of natural gas is often linked to the cost of oil but this equation does not always hold good and there are wide regional variations, depending upon the source of the gas. This is particularly striking in the United States where the development of shale gas reserves during the 21st century has had a dramatic effect, reducing the price of natural gas in that country and making it the fuel of choice for many types of power plant including reciprocating engines. Natural gas in the United States was trading in 2016 at less than one-third of its price in 2008. The price of natural gas in the United States is also much lower than in most other regions of the world, though all have seen the price fall over the past 8−10 years. As with oil, it is therefore important to take account of the risk associated with price fluctuations when planning generating capacity based on natural gas.

CAPITAL COSTS

The capital cost of an engine is also an important part of the LCOE equation. In some cases it can be the sole determining factor when a decision about power generating technology is made. The lowest cost engines available are small petrol-driven devices based on car engines which are manufactured in large numbers each year. These engines can be purchased as stand-by generators for as little as $250/kW. Such engines are cheap so they are well suited to applications where they will only be required to operate infrequently. However they are expensive to run since their energy conversion efficiency is relatively low and they have short lifetimes and require the extensive and regular maintenance of an automotive engine. Against that, there are technicians in just about every part of the world who are capable of maintaining such an engine.

Large engines designed for power generation are generally much more expensive. Natural gas-fired engines of around 300 kW are likely to cost around $2000/kW. While detailed cost data tends to be proprietary, evidence suggests that capital costs drop as engine size rises into

Table 9.1 Cost and Efficiency Figures for a Series of Reciprocating Engine Systems

	Capacity (kW)	Total Installed Cost ($/kW)	Engine Speed (rpm)	Efficiency (HHV, %)
System 1	100	2210	1800	28.4
System 2	300	1940	1800	34.6
System 3	800	1640	1800	35.0
System 4	3000	1130	900	36.0
System 5	5000	1130	720	39.0

Source: *US Environmental Protection Agency.*

the megawatt and multi-megawatt range. All these larger engines are built to be able to operate for long periods between maintenance. They are generally more efficient than the smaller engines too, so their operating costs are lower.

Table 9.1 shows figures for a range of generic power generation systems based on reciprocating engines which illustrate these trends. The costs in this table are in 2010 dollars. (Figures from the US Energy Information Administration suggest that costs for engine-based distributed generation systems rose by around 8% between 2010 and 2016.) The smallest system in the table, with generating capacity of 100 kW, is based on a high-speed engine operating at 1800 rpm. Efficiency is 28.4% and the cost is $2210/kW. At the other end of the scale a 5 MW system is based on a medium-speed engine running at 720 rpm. This has an efficiency of 39.0% and an installed cost of $1130/kW.

The cost-effectiveness of most systems such as those in the table above will depend on whether they can be used to supply heat as well as electrical power. All the above systems were assessed for their cogeneration efficiency when providing hot water. The most efficient were systems 2 and 3 with overall efficiencies of 78% and 79%, respectively. The least efficient was system 4 at 73%.

The cost of Stirling engines is much higher than for most internal combustion engines because they are not produced in large enough quantities to bring about the economy of volume production. Estimates vary widely, from as low as $2000/kW to as high as $50,000/ kW. The cost of a domestic Stirling engine CHP system with a generating capacity of 1 kW appears to be around £2000, although accurate figures are scarce.

THE LEVELIZED COST OF ELECTRICITY

Figures for the LCOE from reciprocating engines are also scarce. However Lazard has recently published some estimates for the cost in the United States.[2] The figures are for reciprocating engines used for distributed generation. The analysis found that the LCOE for a diesel reciprocating engine was between $212/MWh and $281/MWh. The comparable cost for a natural gas reciprocating engine was from $68/MWh to $101/MWh. In both cases the lower figure is for a unit used for base-load generation while the higher is for backup or intermittent use with a 10% duty cycle. The figures suggest that a natural gas engine is one of the cheapest sources of electricity based on fossil fuel. However the low cost of electricity relies substantially on the low cost of natural gas in the United States. Costs elsewhere are likely to be higher.

[2]Lazard's Levelized Cost of Energy Analysis – Version 10.0, December 2016.

INDEX

Printed in the United States
By Bookmasters